KUBOK®

KUBOK COLLECTION n.1
160 puzzles

LOGIC PUZZLES
MULTILEVEL
FOR EVERYONE

26

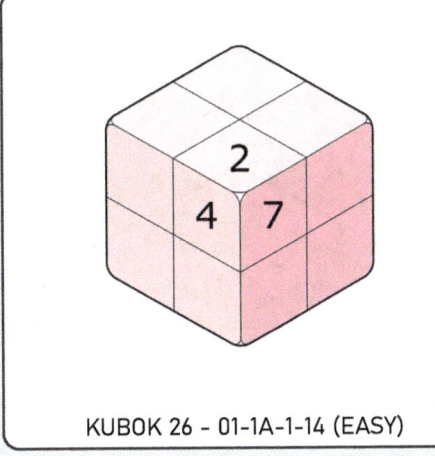

KUBOK 26 - 01-1A-1-14 (EASY)

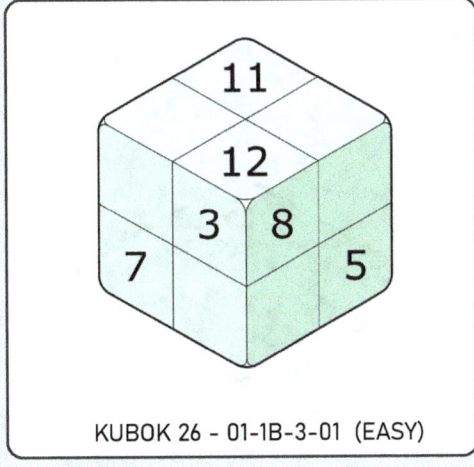

KUBOK 26 - 01-1B-3-01 (EASY)

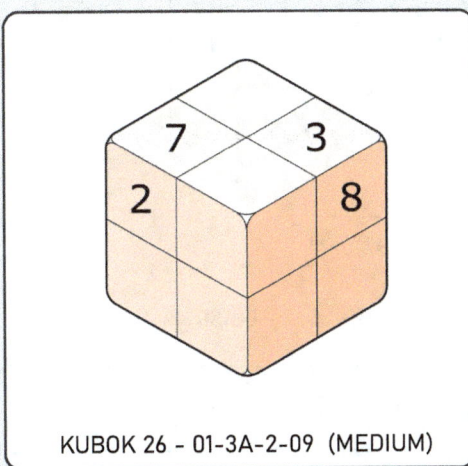

KUBOK 26 - 01-3A-2-09 (MEDIUM)

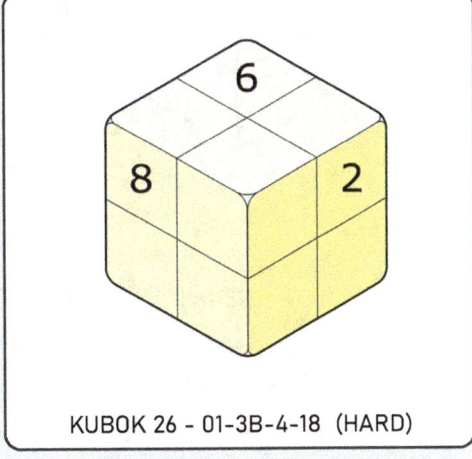

KUBOK 26 - 01-3B-4-18 (HARD)

Fig. A

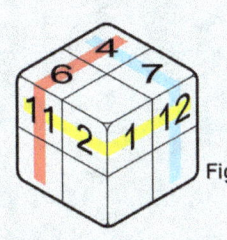

Fig. B

KUBOK 26 - RULES:

In each scheme the numbers from 1 to 12 must be placed without repetitions.

fig. A - The sum of the numbers present in each of the 3 faces of the cube must be equal to 26

fig. B - The sum of the numbers present in each of the 6 rows composed of rows and columns must be equal to 26

www.kubok.it

KUBOK 3x9 - 03180722 (MEDIUM)

KUBOK 3x9 - 08180719 (HARD)

KUBOK 3x9 - RULES:

fig. A

fig. B

fig. C

fig. A - In each of the 3 faces of the cube they must be present all numbers from 1 to 9 without repetitions.

fig. B, C - The sum of the six numbers, all different between them, present in each of the 9 rows composed of rows and columns, must be equal to 30

www.kubok.it

KUBOK 8 - 05190621 - (EASY)

KUBOK 8 - 05190615 - (MEDIUM)

KUBOK 8 - RULES:

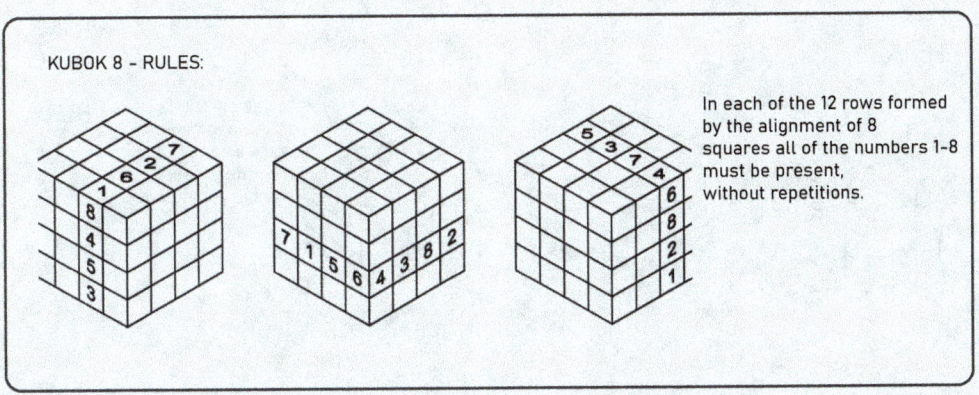

In each of the 12 rows formed by the alignment of 8 squares all of the numbers 1–8 must be present, without repetitions.

www.kubok.it

KUBOK 8 - 04190609 (HARD)

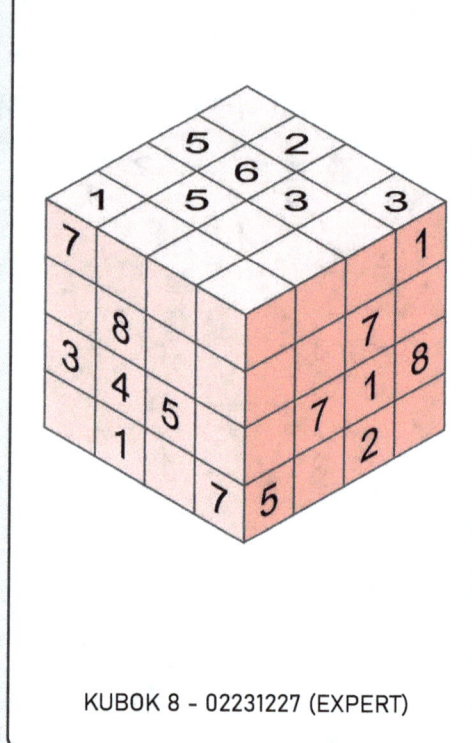

KUBOK 8 - 02231227 (EXPERT)

KUBOK 8 - RULES:

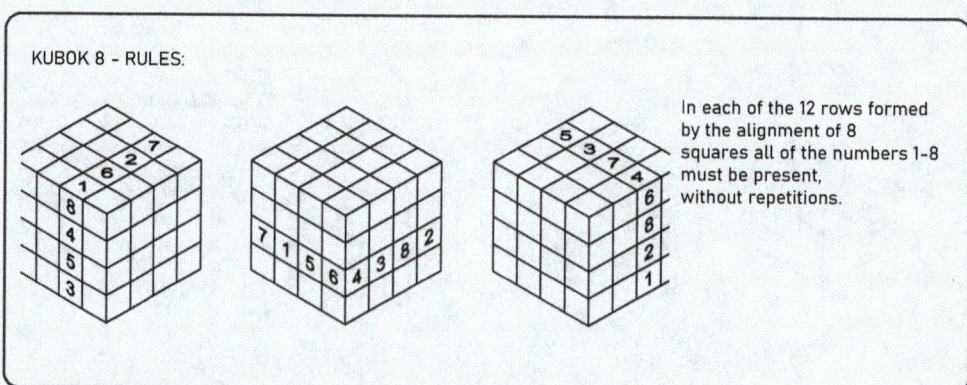

In each of the 12 rows formed by the alignment of 8 squares all of the numbers 1–8 must be present, without repetitions.

www.kubok.it

KUBOK® 12

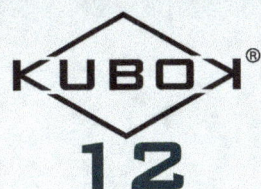

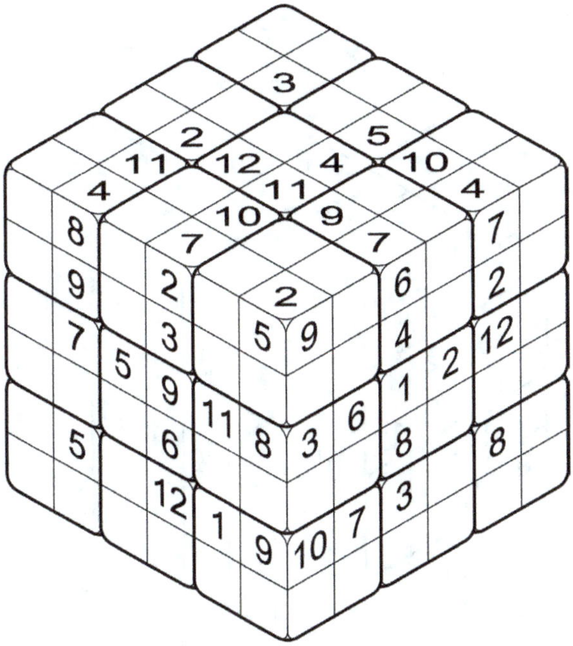

KUBOK 12 - 004205 (EASY)

KUBOK 12 - RULES:
In each of the 18 rows formed by the alignment of 12 squares all of the numbers from 1 to 12 must be present, without any repetition; (A)
In each of the 6 zones made up of 6 x 2 squares, both horizontal and vertical, identifiable on each face of the cube, all of the numbers from 1 to 12 must be present, without any repetition; (Fig. B)
All of the numbers from 1 to 12 must be present without any repetition in the shape formed by the cube's three faces that form the central corner of 2 x 2 x 2; (Fig. C)
in the shape formed by two visible faces of the six 2 x 2 x 2 cubes on the edges of the large cube, there must be only 8 of the 12 numbers from 1 to 12 present, without any repetition; (Fig. D)

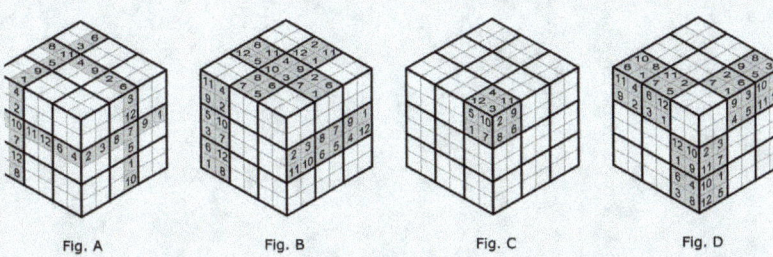

Fig. A

Fig. B

Fig. C

Fig. D

KUBOK 12 - 003122 (MEDIUM)

KUBOK 12 - RULES:

In each of the 18 rows formed by the alignment of 12 squares all of the numbers from 1 to 12 must be present, without any repetition; (A)

In each of the 6 zones made up of 6 x 2 squares, both horizontal and vertical, identifiable on each face of the cube, all of the numbers from 1 to 12 must be present, without any repetition; (Fig. B)

All of the numbers from 1 to 12 must be present without any repetition in the shape formed by the cube's three faces that form the central corner of 2 x 2 x 2; (Fig. C)

in the shape formed by two visible faces of the six 2 x 2 x 2 cubes on the edges of the large cube, there must be only 8 of the 12 numbers from 1 to 12 present, without any repetition; (Fig. D)

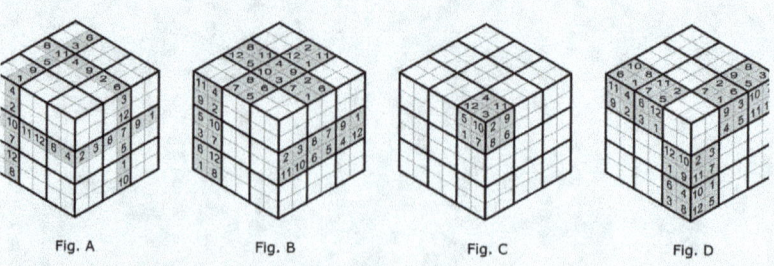

Fig. A   Fig. B   Fig. C   Fig. D

www.kubok.it

KUBOK 12 - 09031301 (HARD)

KUBOK 12 - RULES:

In each of the 18 rows formed by the alignment of 12 squares all of the numbers from 1 to 12 must be present, without any repetition; (A)

In each of the 6 zones made up of 6 x 2 squares, both horizontal and vertical, identifiable on each face of the cube, all of the numbers from 1 to 12 must be present, without any repetition; (Fig. B)

All of the numbers from 1 to 12 must be present without any repetition in the shape formed by the cube's three faces that form the central corner of 2 x 2 x 2; (Fig. C)

in the shape formed by two visible faces of the six 2 x 2 x 2 cubes on the edges of the large cube, there must be only 8 of the 12 numbers from 1 to 12 present, without any repetition; (Fig. D)

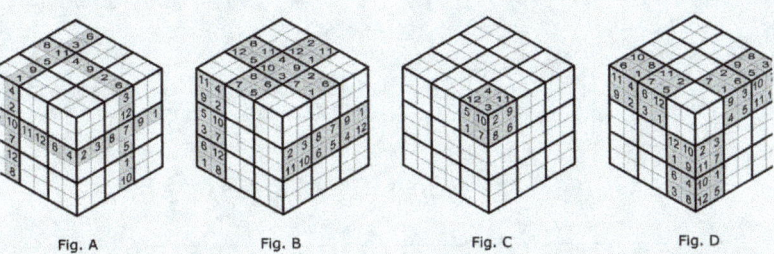

Fig. A   Fig. B   Fig. C   Fig. D

www.kubok.it

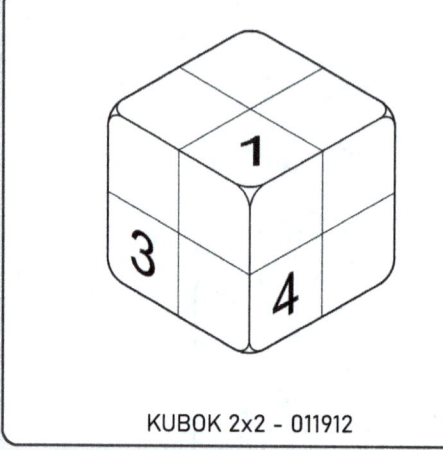

KUBOK 2x2 - 011912

KUBOK 2x2 - 011913

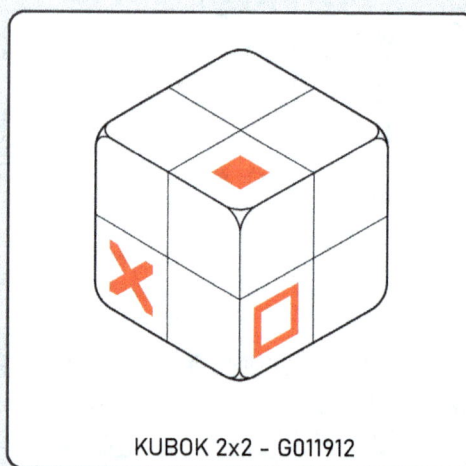

KUBOK 2x2 - G011912

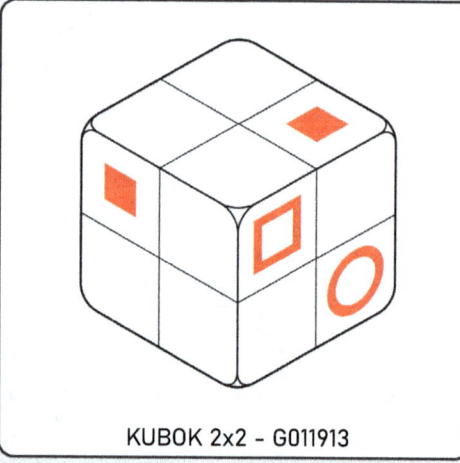

KUBOK 2x2 - G011913

fig.A

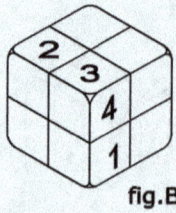

fig.B

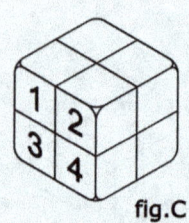

fig.C

KUBOK 2x2 - RULES:

fig. A – C   Fill all the boxes so that each row, each column and every face of the 2x2 box contains all the numbers 1–4, or the four graphic symbols, without repetitions.

www.kubok.it

EQUI

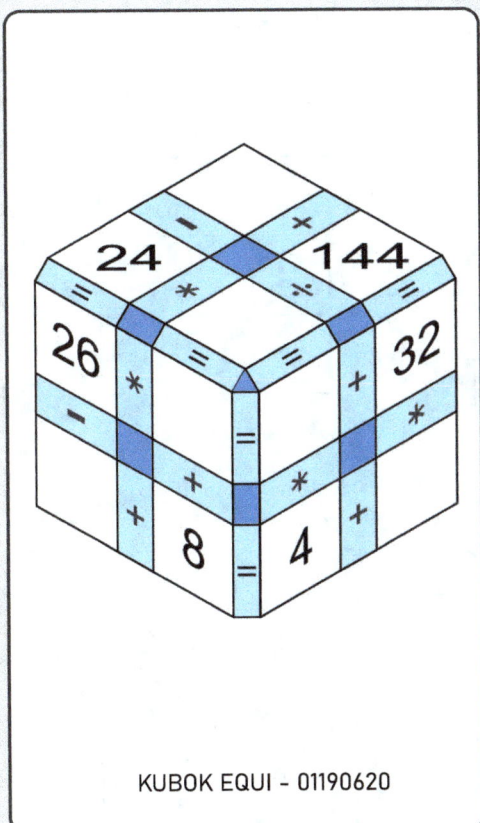

KUBOK EQUI - 01190620

KUBOK EQUI - 03160930

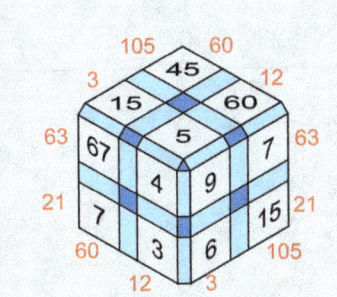

KUBOK EQUI - RULES:

In each of scheme there are 6 arithmetic equations; insert the missing numbers in the empty boxes in order to satisfy all the equations.

www.kubok.it

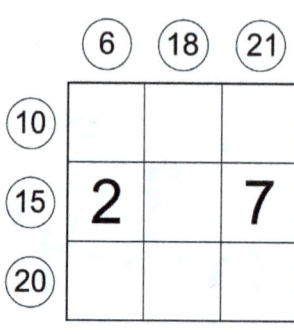

K9 - 100514 (EASY)

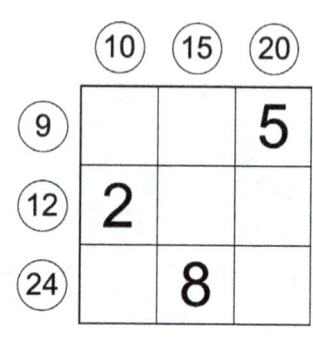

K9 - 100916 (EASY)

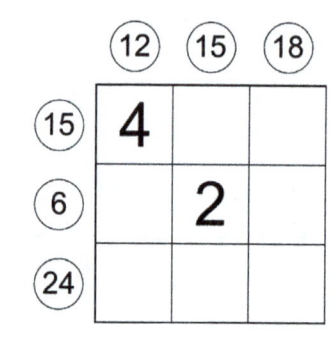

K9 - 100421 (MEDIUM)

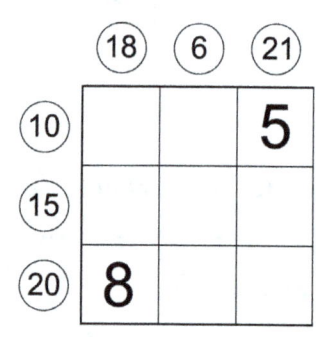

K9 - 100623 (MEDIUM)

KUBOK 9 - RULES:

Enter the missing numbers 1-9 without repetitions so that the sum of the three numbers in each row and column is the same as the corresponding circled number

www.kubok.it

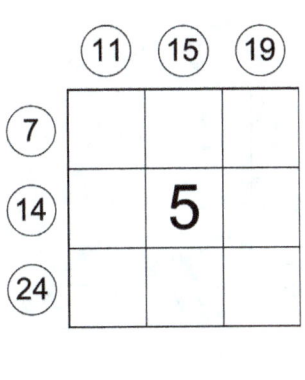

K9 – 100331 (HARD)

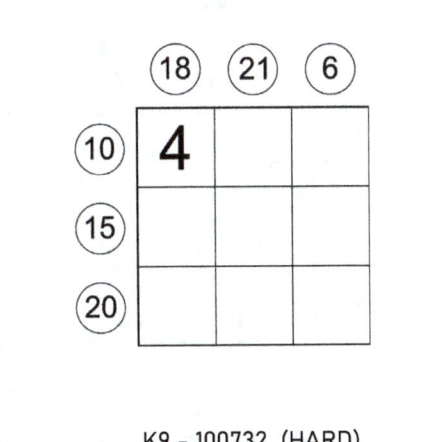

K9 – 100732 (HARD)

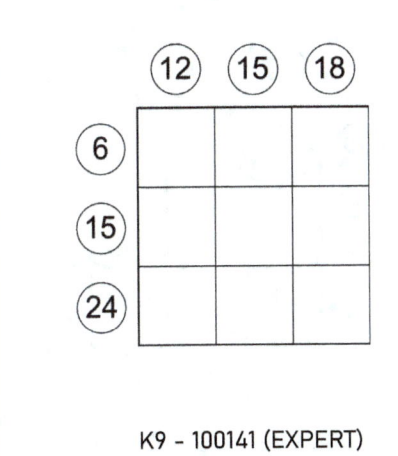

K9 – 100141 (EXPERT)

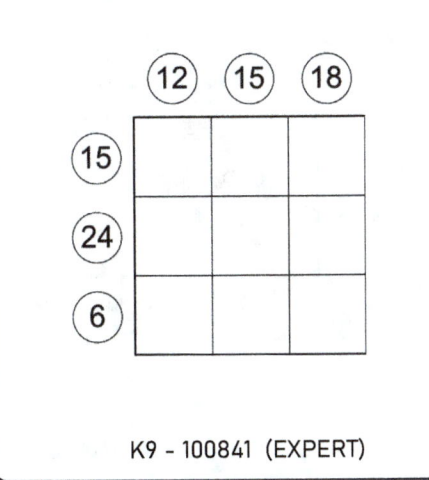

K9 – 100841 (EXPERT)

KUBOK 9 - RULES:

Enter the missing numbers 1-9 without repetitions so that the sum of the three numbers in each row and column is the same as the corresponding circled number

www.kubok.it

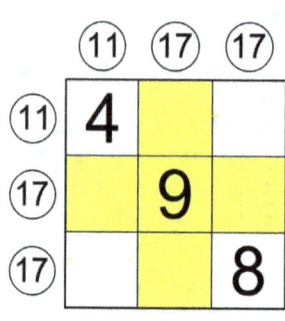

K9 ODD EVEN – 202301 (EASY)

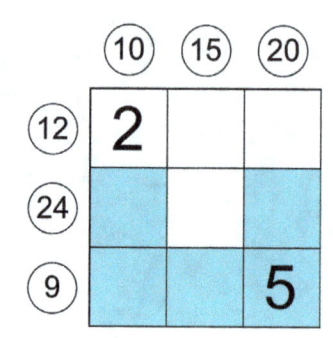

K9 ODD EVEN – 111422 (MEDIUM)

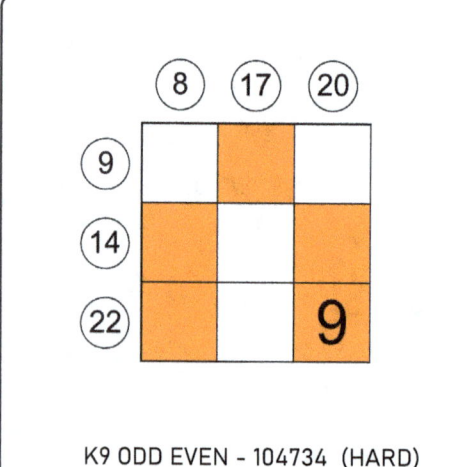

K9 ODD EVEN – 104734 (HARD)

K9 ODD EVEN – 105645 (EXPERT)

KUBOK 9 ODD EVEN - RULES:

Enter the missing numbers 1–9 without repetitions so that the sum of the three numbers in each row and column is the same as the corresponding circled number. Even numbers will only be placed in the white boxes.

www.kubok.it

16

K16 - 100111 - EASY

	(26)	(34)	(37)	(39)
(48)				15
(17)	6	2	4	
(33)		14	11	7
(38)	3			

K16 - 100512 - EASY

	(35)	(47)	(34)	(20)
(36)		15		3
(22)		5	14	
(42)		16	6	
(36)	12		4	

K16 - 100221 - MEDIUM

	(30)	(37)	(40)	(29)
(38)	6			1
(41)		9	12	
(19)		4	2	
(38)	14			5

K16 - 100622 - MEDIUM

	(52)	(20)	(38)	(26)
(22)		1		3
(40)		2		12
(37)	15		9	
(37)	8		14	

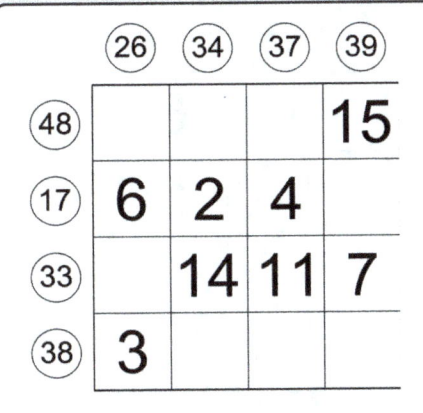

	(19)	(35)	(30)	(52)
(36)			4	
(41)		14	13	
(35)	2	6	12	15
(24)			1	

KUBOK 16 - RULES:

Enter the missing numbers 1-16 without repetitions so that the sum of the four numbers in each row and column is the same as the corresponding circled number

www.kubok.it

16

K16 - 100331 - HARD

Top: (51) (26) (28) (31)

(26) 8			11
(36)		9	
(24)	5		
(50) 16			

K16 - 100735 - HARD

Top: (37) (28) (32) (39)

(41)		16	
(48)		8	13
(23) 11	3		
(24)	9		

K16 - 100441 - EXPERT

Top: (24) (28) (41) (43)

(48)			13
(25)		14	
(44)	10		
(19) 5			

K16 - 100842 - EXPERT

Top: (33) (36) (47) (20)

(30)			13
(36)	11		
(28)		12	
(42) 15			

Top: (19) (35) (30) (52)

(36)		4	
(41) 14	13		
(35) 2	6	12	15
(24)	1		

KUBOK 16 - RULES:

Enter the missing numbers 1-16 without repetitions so that the sum of the four numbers in each row and column is the same as the corresponding circled number

16

www.kubok.it

16 ODD EVEN

K16 ODD EVEN - 01231103 - EASY

K16 ODD EVEN - 02231102 - MEDIUM

K16 ODD EVEN - 03231101 - HARD

K16 ODD EVEN - 141541 - EXPERT

KUBOK 16 ODD EVEN - RULES:

Enter the missing numbers 1-16 without repetitions so that the sum of the four numbers in each row and column is the same as the corresponding circled number. Even numbers will only be placed in the white boxes.

www.kubok.it

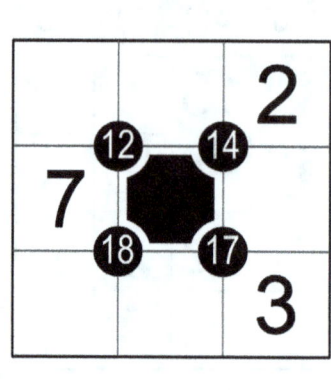

COMPACT 8 - 03231212 - EASY

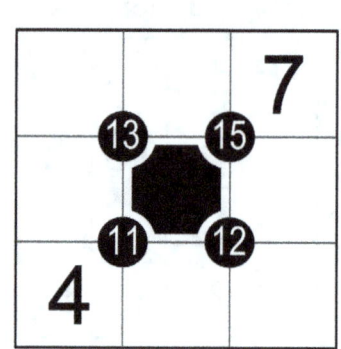

COMPACT 8 - 02231213 - MEDIUM

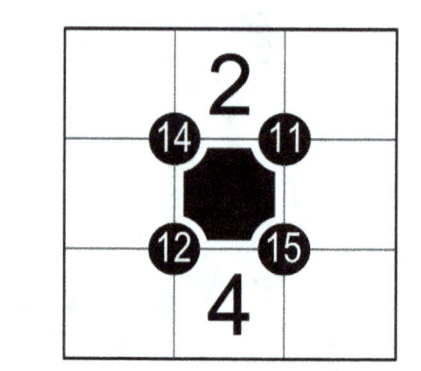

COMPACT 8 - 02231210 - MEDIUM

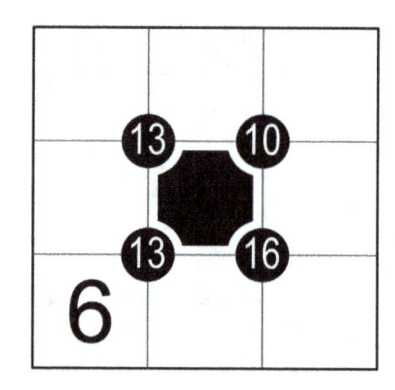

COMPACT 8 - 01231211 - HARD

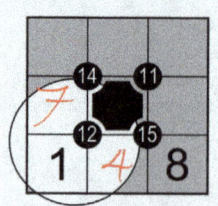

KUBOK COMPACT 8 - RULES:

Enter the missing numbers 1-8 without repetitions so that the sum of the numbers present in each of the 3 white boxes around each small black circle corresponds to the number inside the small circle.

www.kubok.it

COMPACT 12

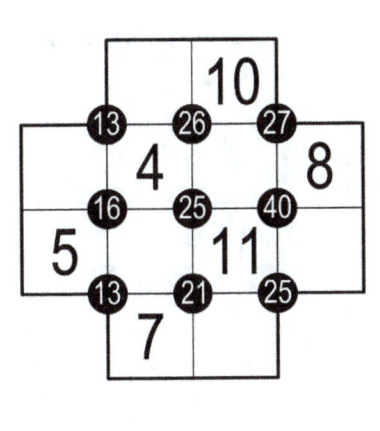

COMPACT 12 - 01231003 - EASY

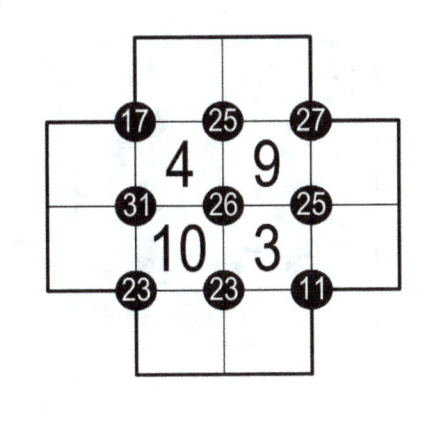

COMPACT 12 - 01231001 - MEDIUM

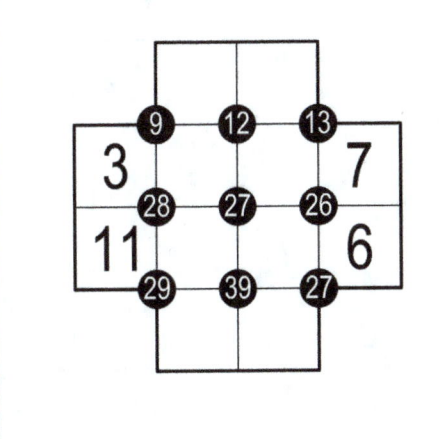

COMPACT 12 - 01231004 - HARD

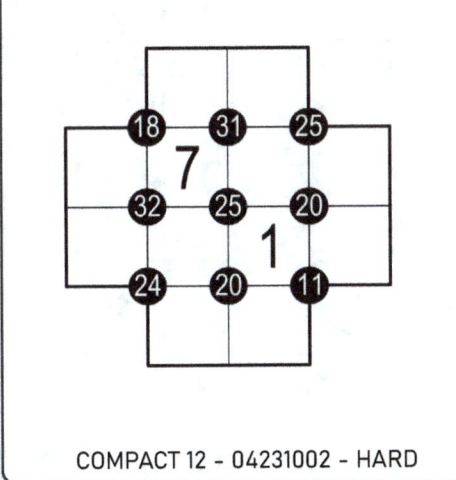

COMPACT 12 - 04231002 - HARD

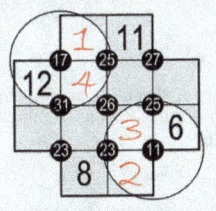

KUBOK COMPACT 12 - RULES:

Insert the missing numbers from 1–12 without repetitions so that the sum of the numbers present in each of the boxes surrounding each small black circle corresponds to the number inside the circle.

www.kubok.it

COMPACT 16

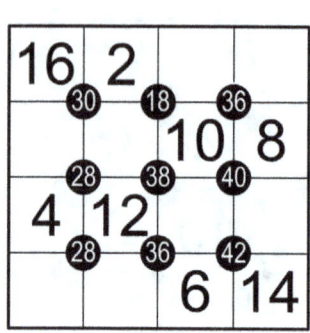

COMPACT 16 – 01231103 – EASY

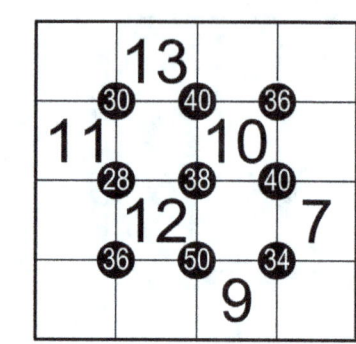

COMPACT 16 – 02231102 – MEDIUM

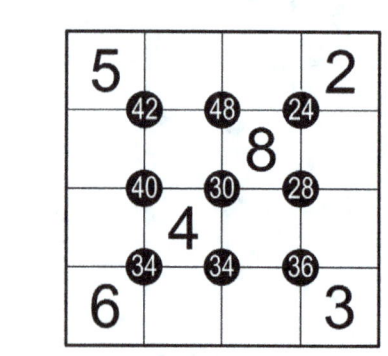

COMPACT 16 – 03231101 – HARD

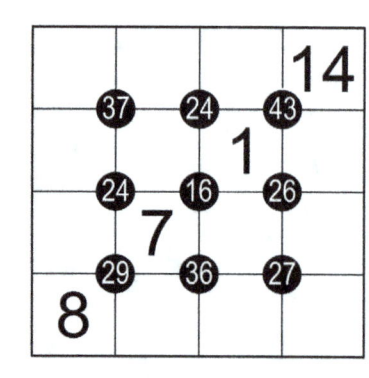

COMPACT 16 – 041415 – EXPERT

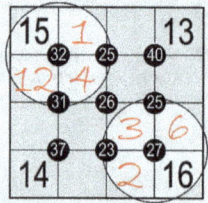

KUBOK COMPACT 16 - RULES:

Fill in the missing numbers 1–16 without repetition so that the sum of the numbers in each of the boxes around each small black circle corresponds to the number inside the small circle.

www.kubok.it

COMPACT 8 ODD EVEN

COMPACT 8 ODD EVEN
01231127 – EASY

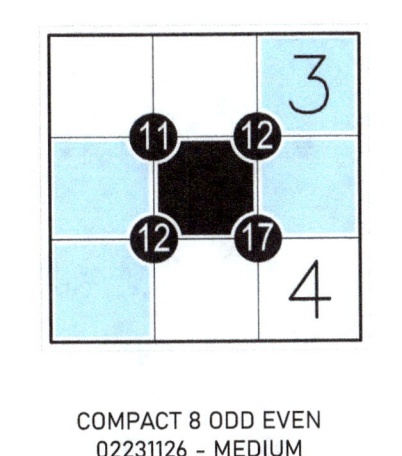

COMPACT 8 ODD EVEN
02231126 – MEDIUM

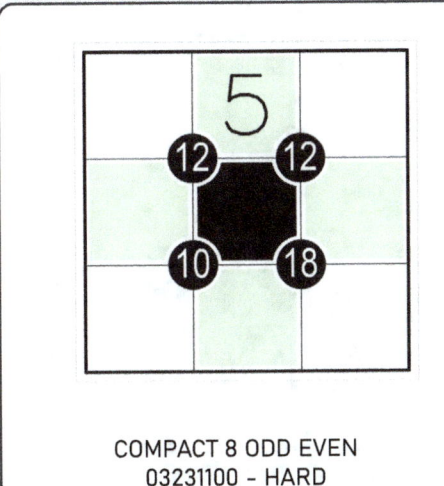

COMPACT 8 ODD EVEN
03231100 – HARD

COMPACT 8 ODD EVEN
04231125 – EXPERT

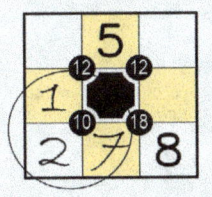

KUBOK COMPACT 8 ODD EVEN - RULES:

Enter the missing numbers 1-8 without repetitions so that the sum of the numbers present in each of the 3 white boxes around each small black circle corresponds to the number inside the small circle. Even numbers will only be placed in the white boxes.

www.kubok.it

COMPACT 9 ODD EVEN

COMPACT 9 ODD EVEN
01231023 - EASY

COMPACT 9 ODD EVEN
02231022 - MEDIUM

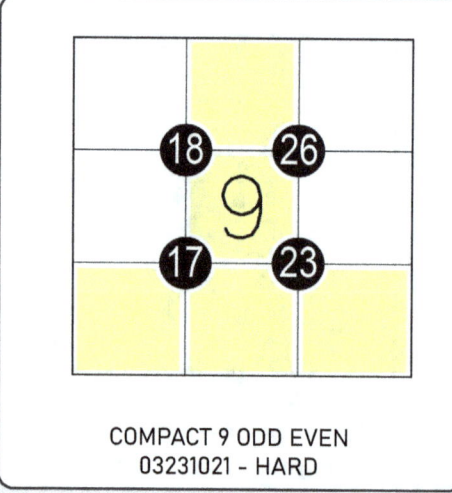

COMPACT 9 ODD EVEN
03231021 - HARD

COMPACT 9 ODD EVEN
04231020 - EXPERT

KUBOK COMPACT 9 ODD EVEN - RULES:

Insert the missing numbers 1–9 without repetitions so that the sum of the numbers present in each of the 4 boxes around each small black circle corresponds to the number inside the circle. Even numbers will only be placed in the white boxes.

www.kubok.it

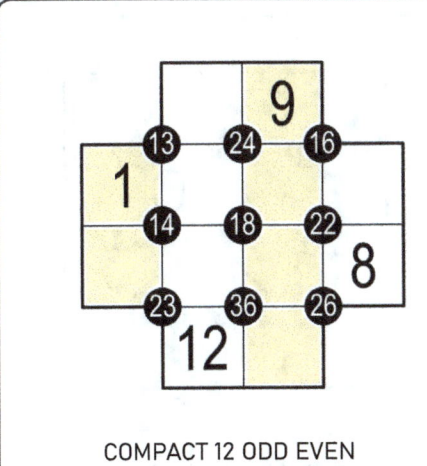

COMPACT 12 ODD EVEN
01231212 - EASY

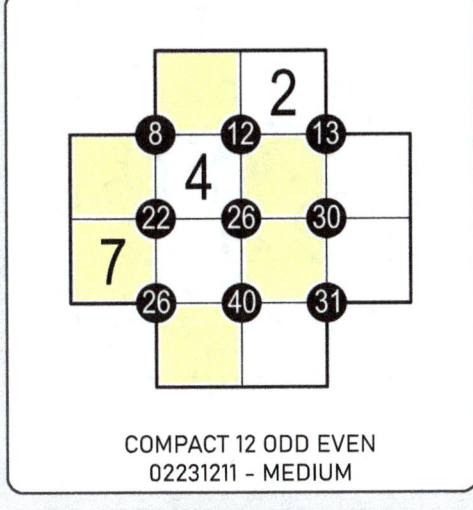

COMPACT 12 ODD EVEN
02231211 - MEDIUM

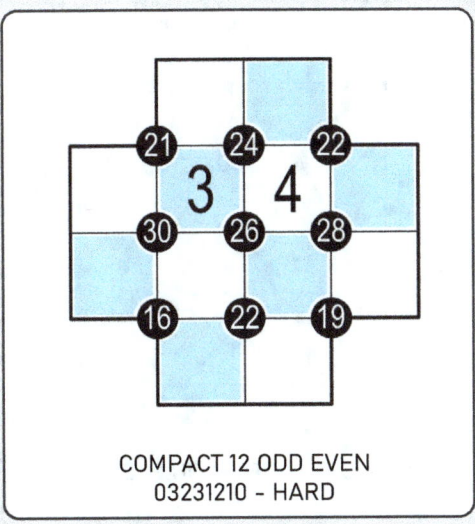

COMPACT 12 ODD EVEN
03231210 - HARD

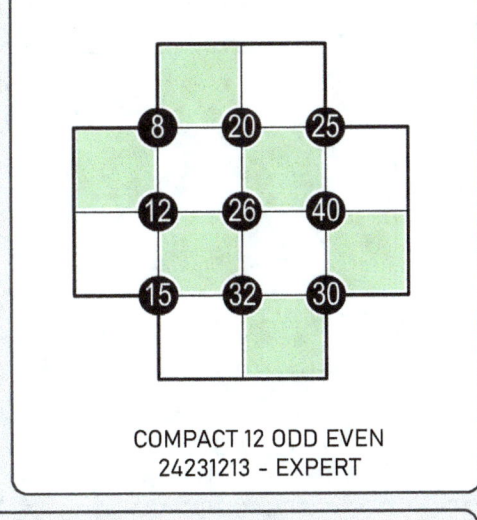

COMPACT 12 ODD EVEN
24231213 - EXPERT

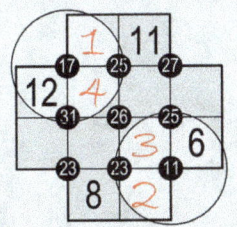

KUBOK COMPACT 12 ODD EVEN - RULES:

Insert the missing numbers 1-12 without repetitions so that the sum of the numbers present in each of the boxes surrounding each small black circle corresponds to the number inside the circle. Even numbers must be entered in the white boxes only.

COMPACT 16 ODD EVEN

COMPACT 16 ODD EVEN
01231103 - EASY

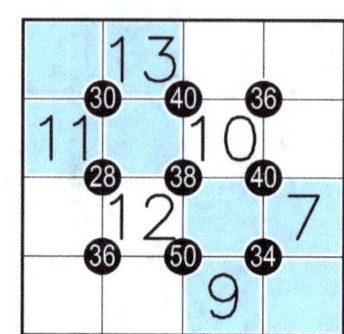

COMPACT 16 ODD EVEN
02231102 - MEDIUM

COMPACT 16 ODD EVEN
00323110 - HARD

COMPACT 16 ODD EVEN
041415 - EXPERT

KUBOK COMPACT 16 ODD EVEN - RULES:

Insert the missing numbers 1–16 without repetitions so that the sum of the numbers present in each of the 4 boxes around each small black circle corresponds to the number inside the circle. Even numbers must be entered in the white boxes only.

www.kubok.it

PYR 4

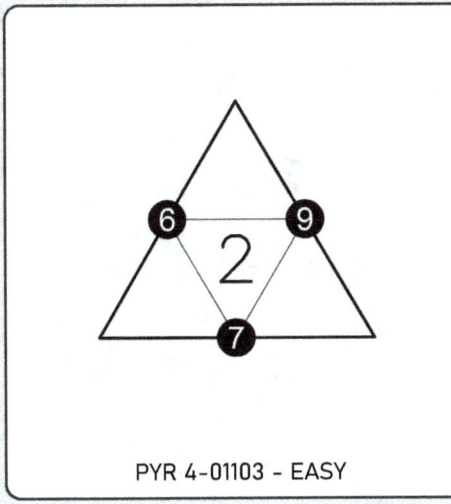

PYR 4-01103 - EASY

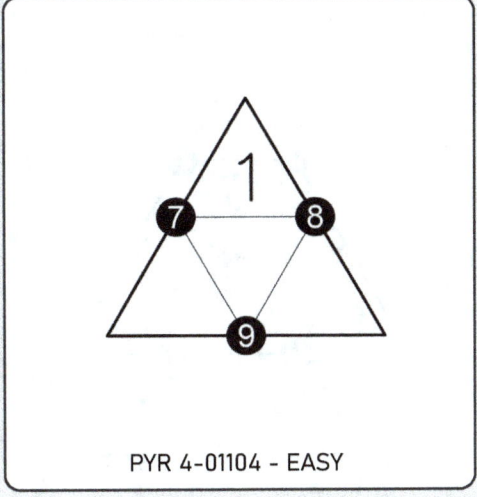

PYR 4-01104 - EASY

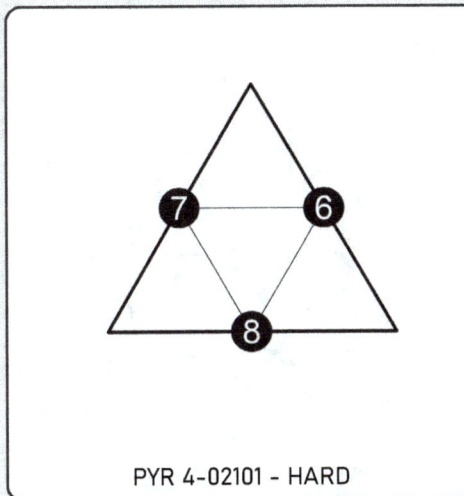

PYR 4-02101 - HARD

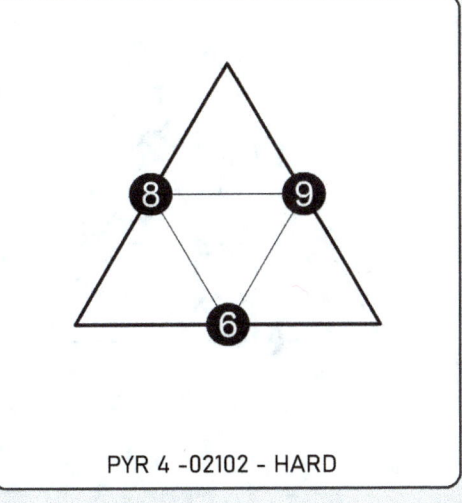

PYR 4 -02102 - HARD

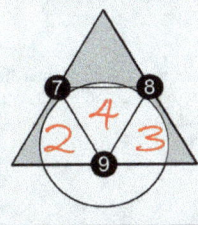

KUBOK PYR 4 - RULES:

Enter the missing numbers 1-4 without repetitions so that the sum of the numbers present in each of the boxes around each black circle corresponds to the number inside the circle.

www.kubok.it

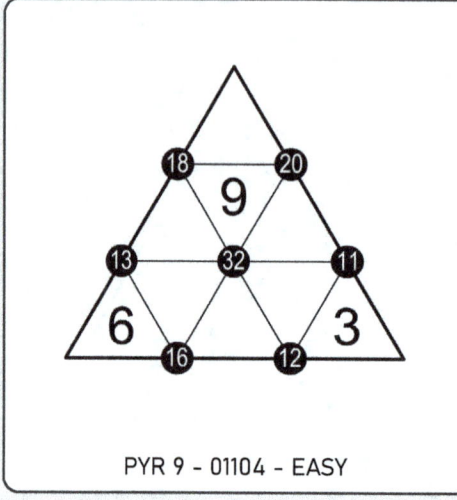

PYR 9 - 01104 - EASY

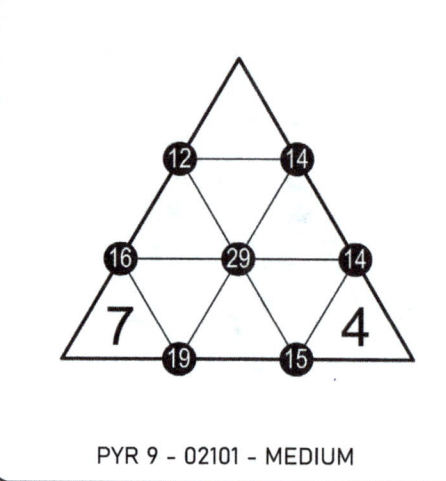

PYR 9 - 02101 - MEDIUM

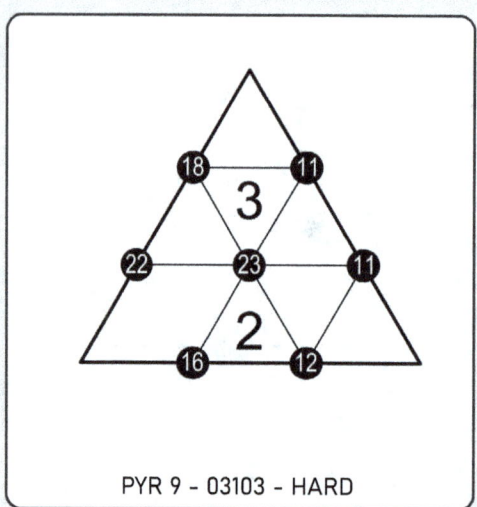

PYR 9 - 03103 - HARD

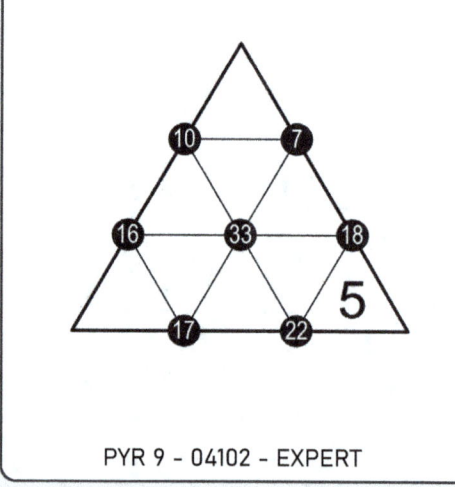

PYR 9 - 04102 - EXPERT

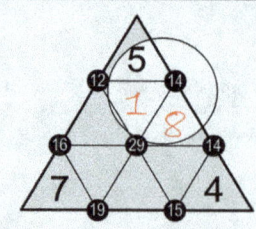

KUBOK PYR 9 - RULES:

Fill in the missing numbers from 1–9 without repetition so that the sum of the numbers in each of the boxes around each small black circle corresponds to the number inside the small circle.

www.kubok.it

PYR 16

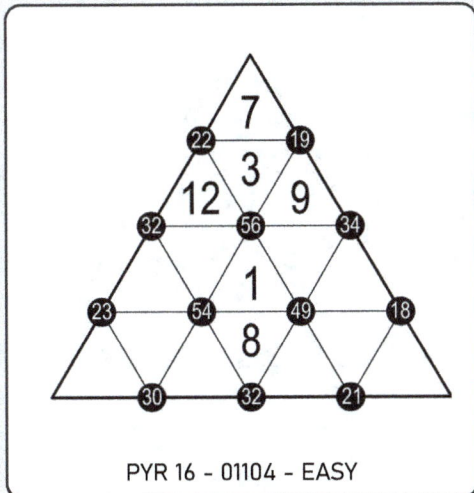

PYR 16 – 01104 – EASY

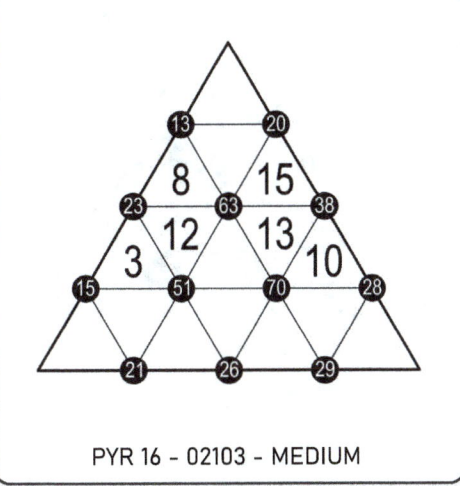

PYR 16 – 02103 – MEDIUM

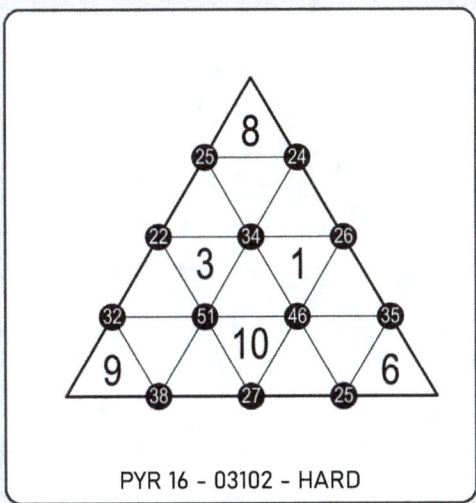

PYR 16 – 03102 – HARD

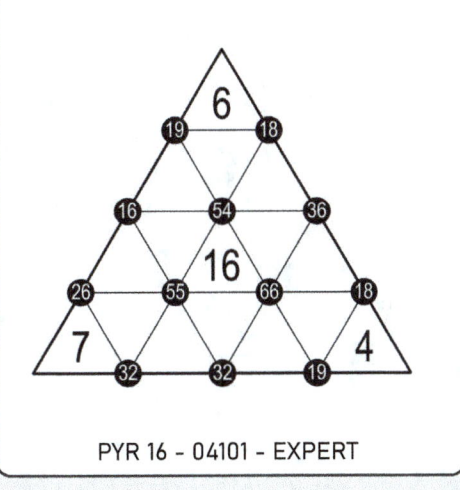

PYR 16 – 04101 – EXPERT

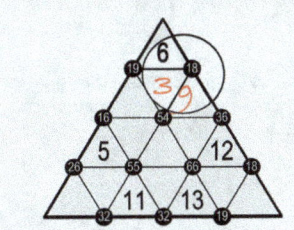

KUBOK PYR 16 – RULES:

Fill in the missing numbers 1-16 without repetition so that the sum of the numbers in each of the boxes around each small black circle corresponds to the number inside the small circle.

www.kubok.it

PILE 6

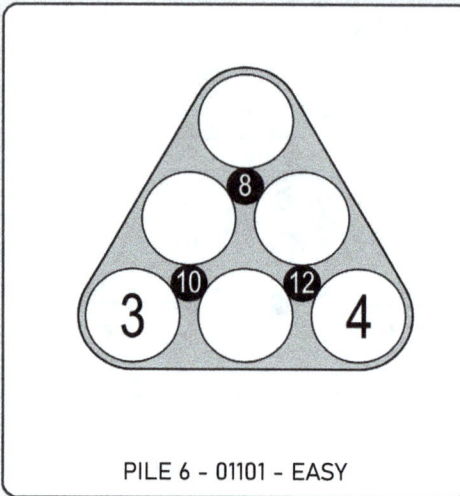

PILE 6 - 01101 - EASY

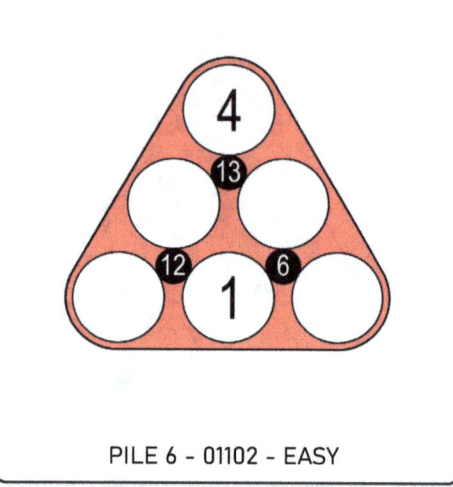

PILE 6 - 01102 - EASY

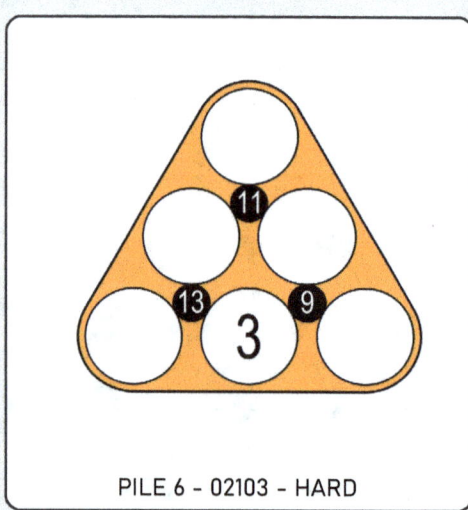

PILE 6 - 02103 - HARD

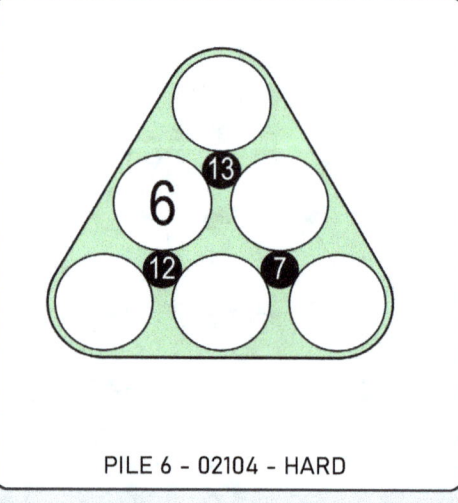

PILE 6 - 02104 - HARD

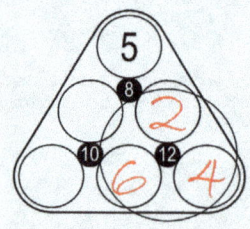

KUBOK PILE 6 - RULES:

Enter the missing numbers 1-6 without repetitions so that the sum of the numbers present in each of the 3 boxes around each small black circle corresponds to the number inside the small circle.

www.kubok.it

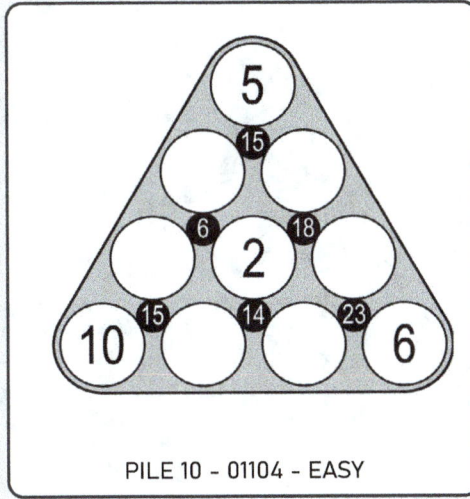

PILE 10 - 01104 - EASY

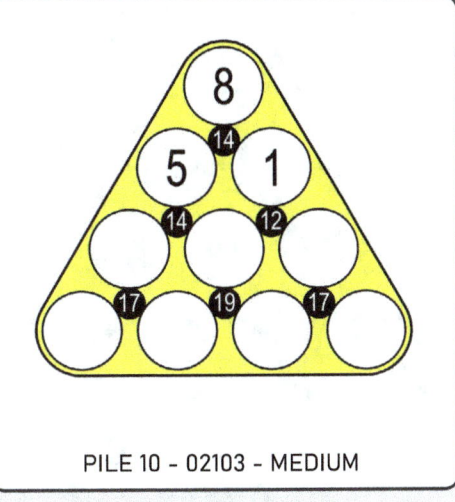

PILE 10 - 02103 - MEDIUM

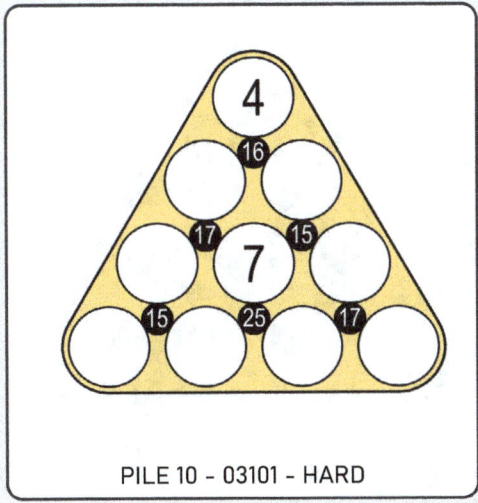

PILE 10 - 03101 - HARD

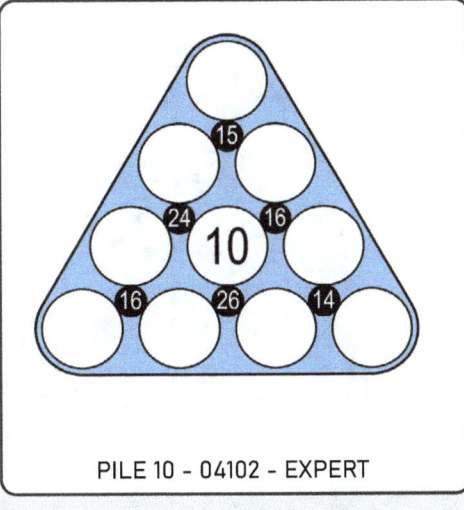

PILE 10 - 04102 - EXPERT

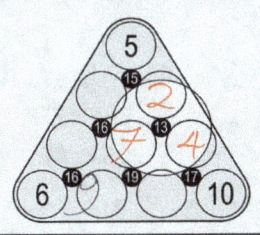

KUBOK PILE 10 - RULES:

Enter the missing numbers 1-10 without repetitions so that the sum of the numbers present in each of the 3 boxes around each small black circle corresponds to the number inside the small circle.

www.kubok.it

PILE 15

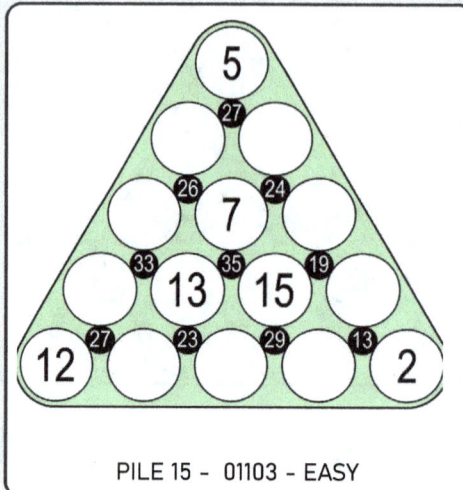

PILE 15 - 01103 - EASY

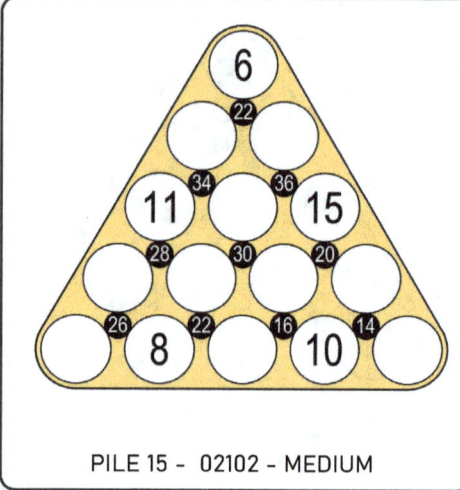

PILE 15 - 02102 - MEDIUM

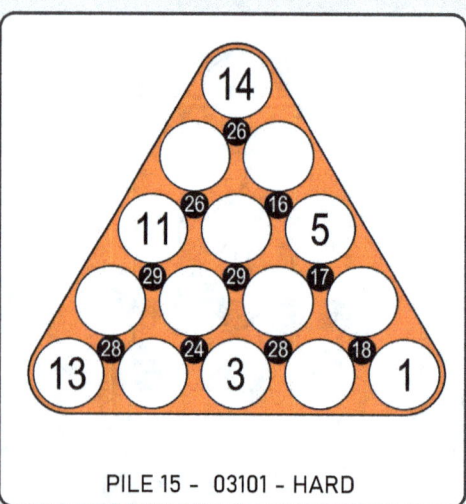

PILE 15 - 03101 - HARD

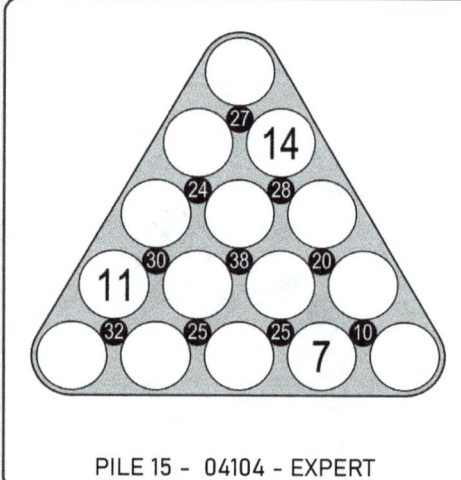

PILE 15 - 04104 - EXPERT

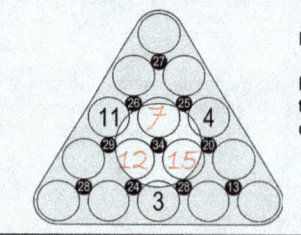

KUBOK PILE 15 - RULES:

Enter the missing numbers 1-15 without repetitions so that the sum of the numbers present in each of the 3 boxes around each small black circle corresponds to the number inside the small circle.

www.kubok.it

PILE 6 ODD EVEN

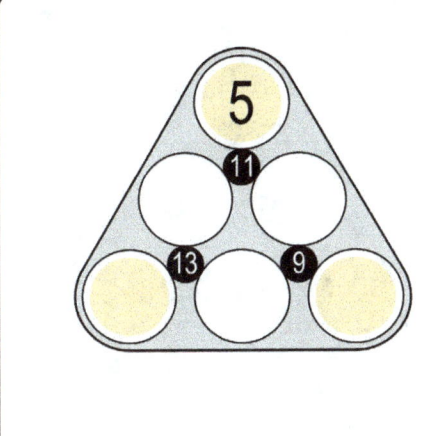

PILE 6 ODD-EVEN - 01101 - EASY

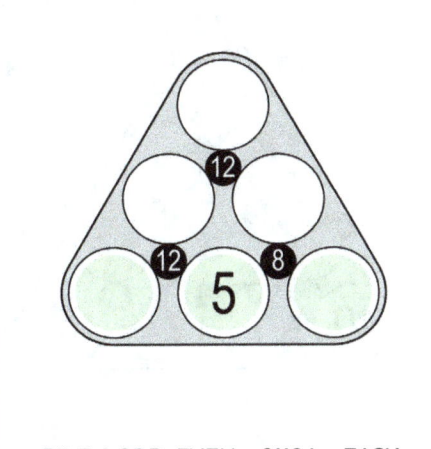

PILE 6 ODD-EVEN - 01104 - EASY

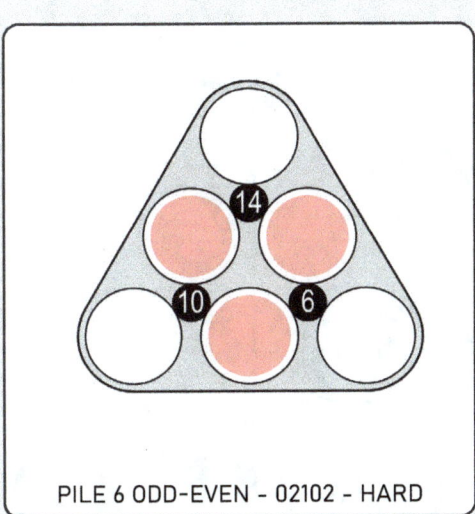

PILE 6 ODD-EVEN - 02102 - HARD

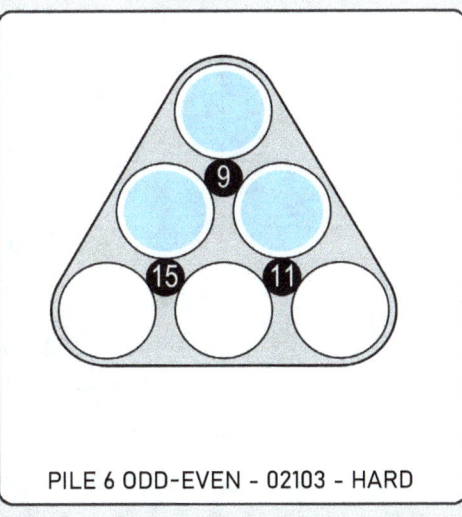

PILE 6 ODD-EVEN - 02103 - HARD

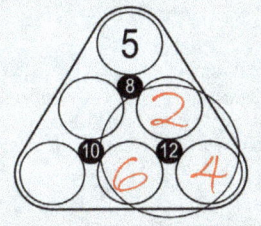

KUBOK PILE 6 ODD EVEN - RULES:

Enter the missing numbers 1-6 without repetitions so that the sum of the numbers present in each of the 3 boxes around each small black circle corresponds to the number inside the small circle. Even numbers will only be placed in the white boxes.

www.kubok.it

PILE 10 ODD EVEN

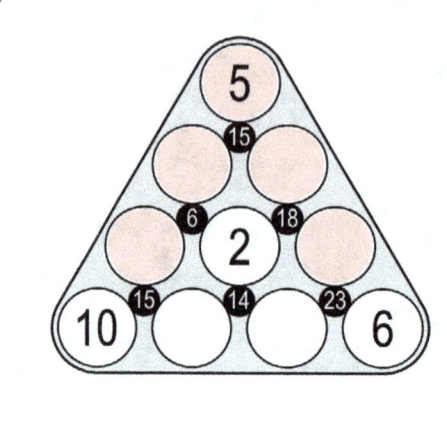

PILE 10 Odd-Even - 01104 - EASY

PILE 10 Odd-Even - 02103 - MEDIUM

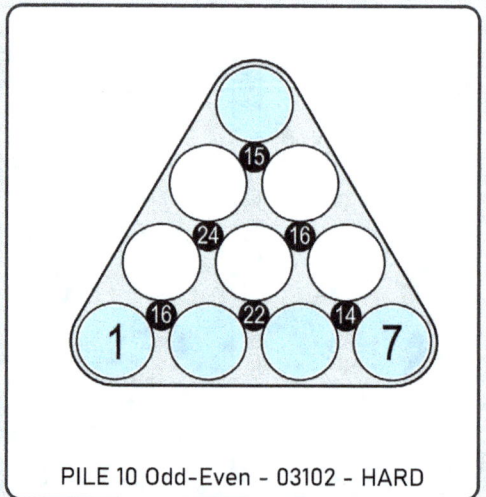

PILE 10 Odd-Even - 03102 - HARD

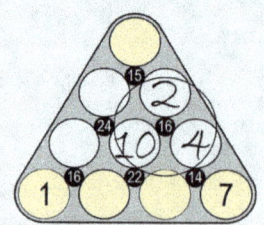

KUBOK PILE 6 ODD EVEN - RULES:

Enter the missing numbers 1-6 without repetitions so that the sum of the numbers present in each of the 3 boxes around each small black circle corresponds to the number inside the small circle. Even numbers will only be placed in the white boxes.

www.kubok.it

PILE 15 ODD EVEN

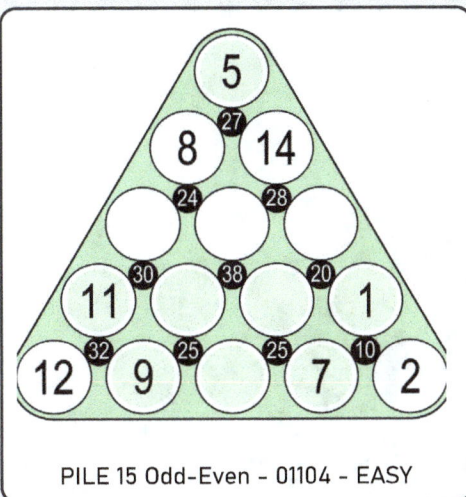

PILE 15 Odd-Even - 01104 - EASY

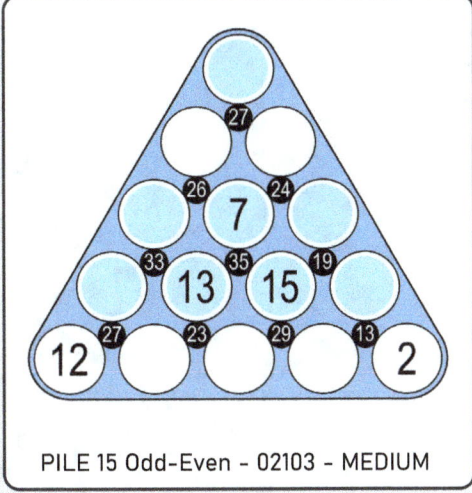

PILE 15 Odd-Even - 02103 - MEDIUM

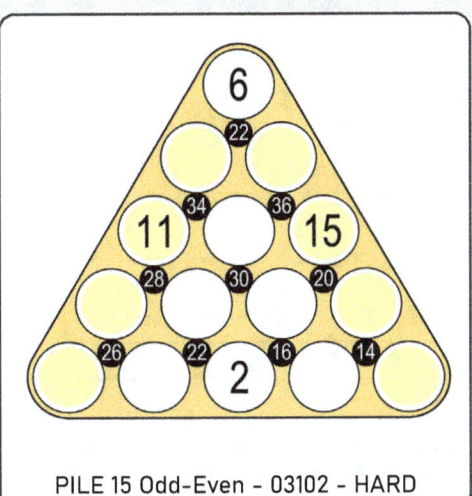

PILE 15 Odd-Even - 03102 - HARD

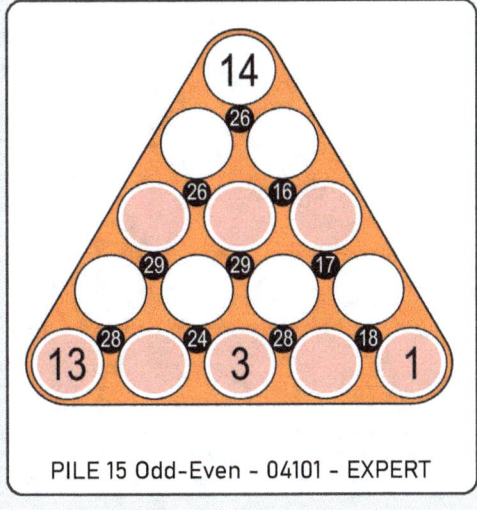

PILE 15 Odd-Even - 04101 - EXPERT

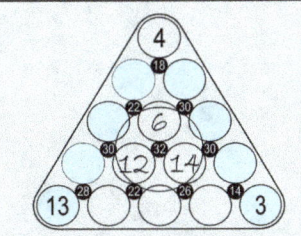

KUBOK PILE 15 ODD EVEN - RULES:

Enter the missing numbers 1-15 without repetitions so that the sum of the numbers present in each of the 3 boxes around each small black circle corresponds to the number inside the small circle. Even numbers will only be placed in the white boxes.

www.kubok.it

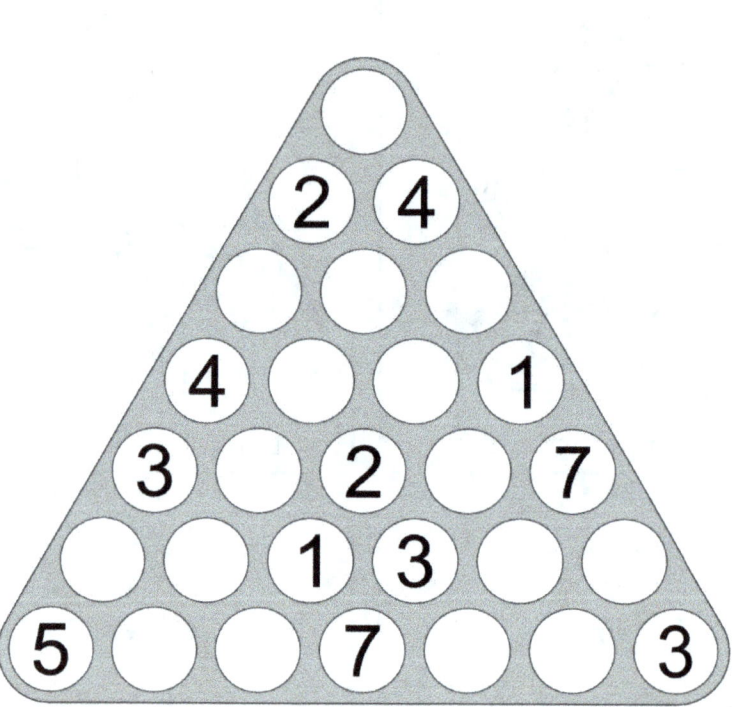

LINK7 01101 - EASY

KUBOK LINK 7 - RULES:

Up to a maximum of seven numbers between 1 and 7 must be entered in each row, without repetitions. Each circular group of seven numbers consisting of a central number and six peripheral numbers must contain the numbers between 1 and 7 without repetition.

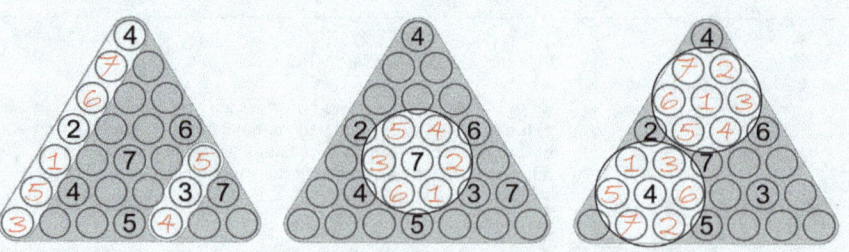

www.kubok.it

LINK 7

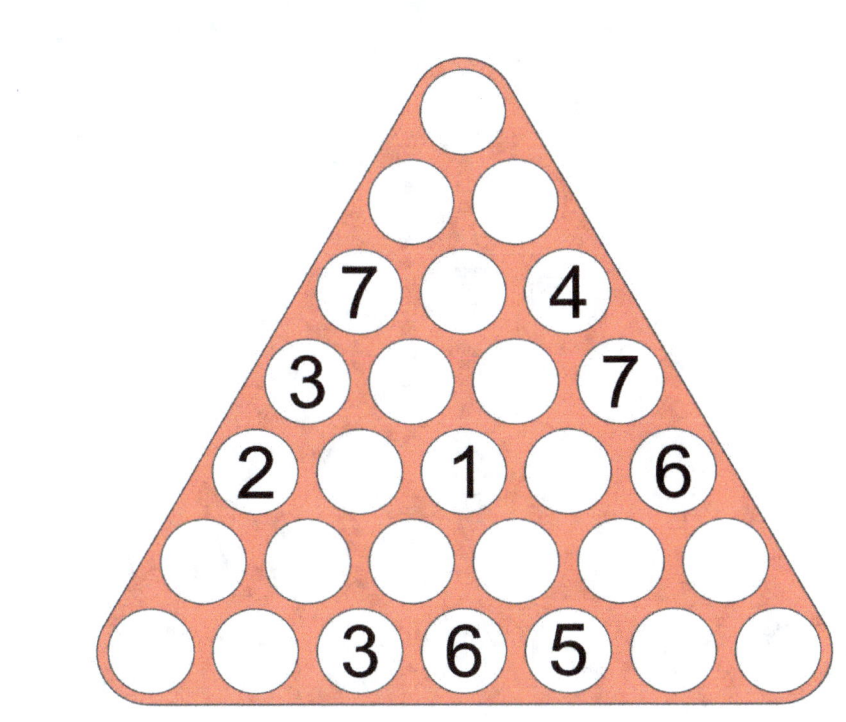

LINK7 02102 - MEDIUM

KUBOK LINK 7 - RULES:

Up to a maximum of seven numbers between 1 and 7 must be entered in each row, without repetitions. Each circular group of seven numbers consisting of a central number and six peripheral numbers must contain the numbers between 1 and 7 without repetition.

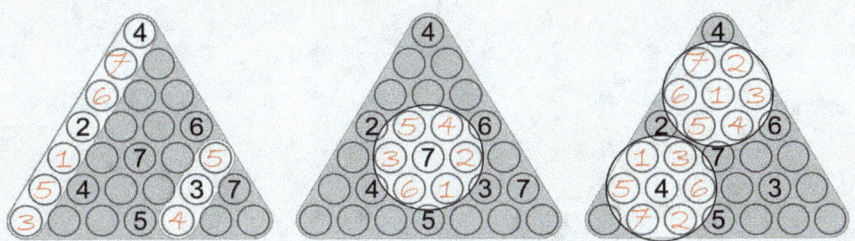

www.kubok.it

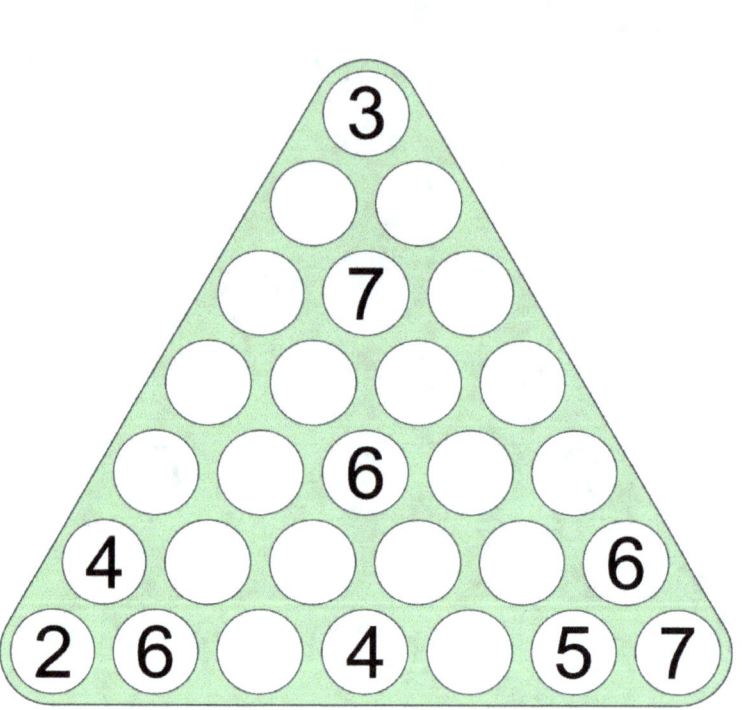

LINK7 03103 - HARD

KUBOK LINK 7 - RULES:

Up to a maximum of seven numbers between 1 and 7 must be entered in each row, without repetitions. Each circular group of seven numbers consisting of a central number and six peripheral numbers must contain the numbers between 1 and 7 without repetition.

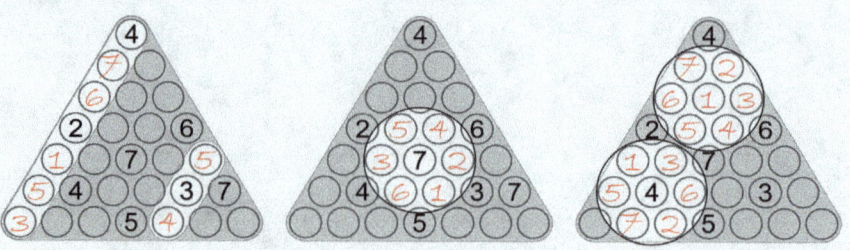

www.kubok.it

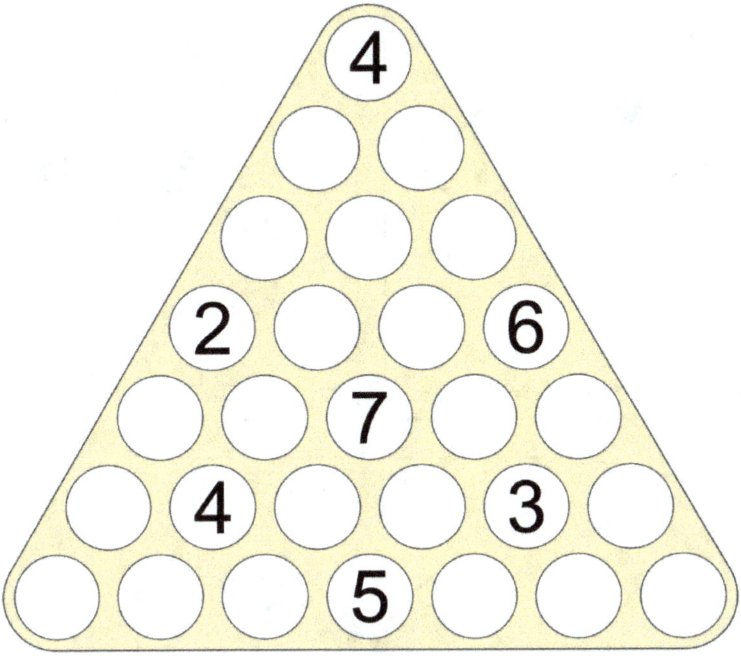

LINK7 04104 - EXPERT

KUBOK LINK 7 - RULES:

Up to a maximum of seven numbers between 1 and 7 must be entered in each row, without repetitions. Each circular group of seven numbers consisting of a central number and six peripheral numbers must contain the numbers between 1 and 7 without repetition.

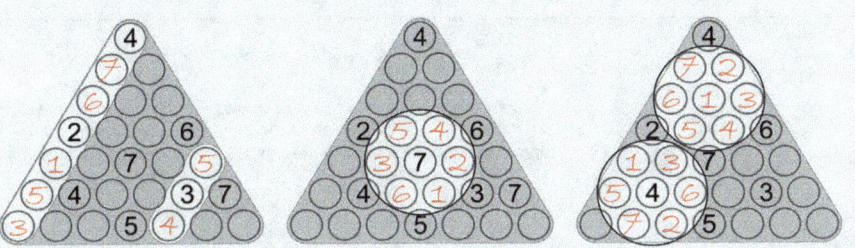

www.kubok.it

ZIG ZAG 5

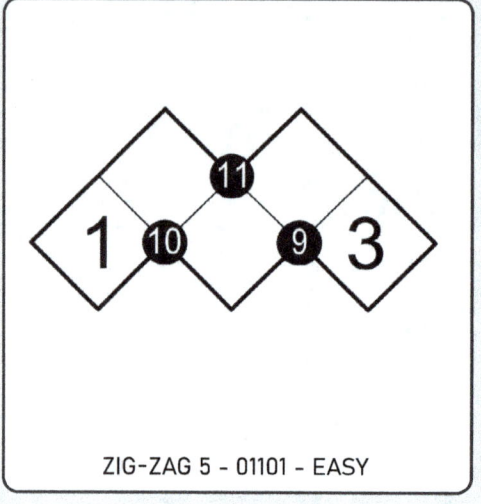

ZIG-ZAG 5 - 01101 - EASY

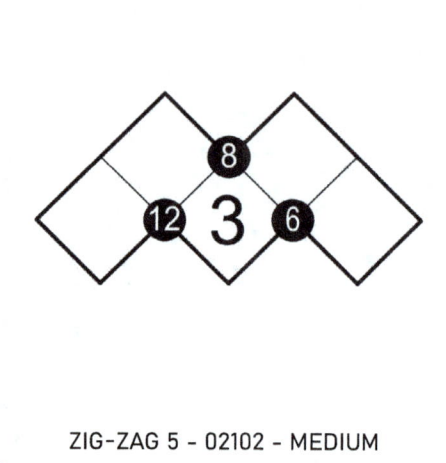

ZIG-ZAG 5 - 02102 - MEDIUM

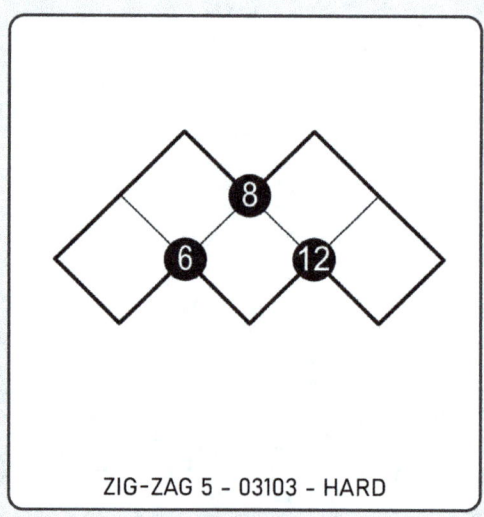

ZIG-ZAG 5 - 03103 - HARD

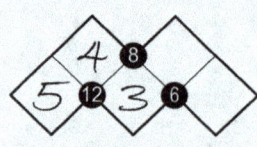

KUBOK ZIG ZAG 5 - RULES:

Enter the missing numbers 1-5 without repetitions so that the sum of the numbers present in each of the 3 boxes around each small black circle corresponds to the number inside the small circle.

www.kubok.it

ZIG ZAG 7

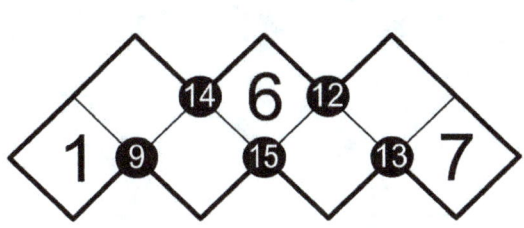

ZIG-ZAG 7 - 01101 - EASY

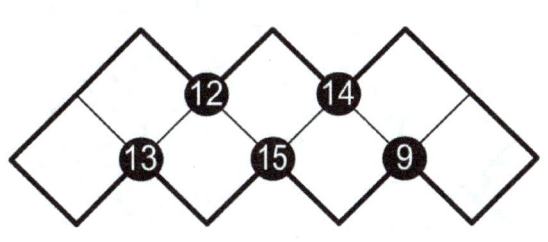

ZIG-ZAG 7 - 04102 - EXPERT

KUBOK ZIG ZAG 7 - RULES:

Enter the missing numbers 1-7 without repetitions so that the sum of the numbers present in each of the 3 boxes around each small black circle corresponds to the number inside the small circle.

www.kubok.it

ZIG ZAG 9

ZIG-ZAG 9 - 01101 - EASY

ZIG-ZAG 9 - 02102 - MEDIUM

ZIG-ZAG 9 - 03103 - HARD

ZIG-ZAG 9 - 04104- EASY

KUBOK ZIG ZAG 9 - RULES:

Enter the missing numbers 1-9 without repetitions so that the sum of the numbers present in each of the 3 boxes around each small black circle corresponds to the number inside the small circle.

www.kubok.it

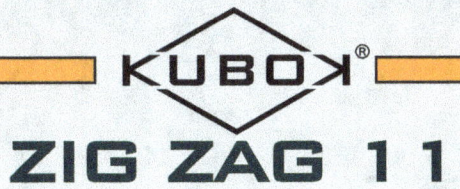

ZIG ZAG 11

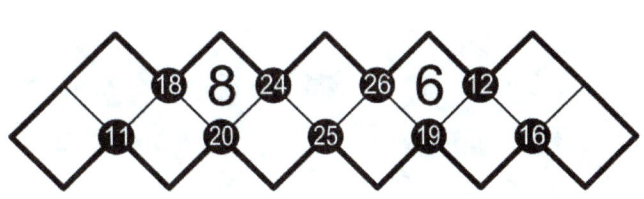

ZIG-ZAG 11 - 02102 - MEDIUM

ZIG-ZAG 11 - 04104 - EXPERT

KUBOK ZIG ZAG 11 - RULES:

Enter the missing numbers 1-11 without repetitions so that the sum of the numbers present in each of the 3 boxes around each small black circle corresponds to the number inside the small circle.

www.kubok.it

ZIG ZAG 13

ZIG-ZAG 13 - 01101 - EASY

ZIG-ZAG 13 - 03103 - HARD

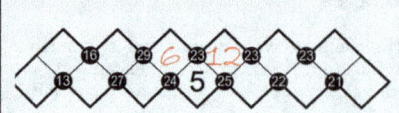

KUBOK ZIG ZAG 13 - RULES:

Enter the missing numbers 1-13 without repetitions so that the sum of the numbers present in each of the 3 boxes around each small black circle corresponds to the number inside the small circle.

www.kubok.it

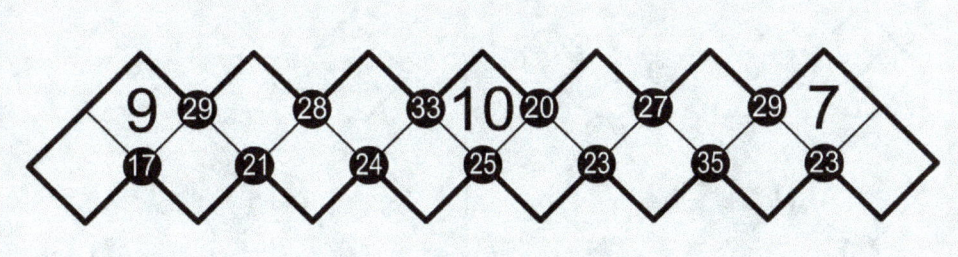

ZIG-ZAG 15 - 02101 - MEDIUM

ZIG-ZAG 15 - 04104 - EXPERT

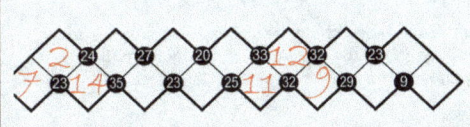

KUBOK ZIG ZAG 15 - RULES:

Enter the missing numbers 1–15 without repetitions so that the sum of the numbers present in each of the 3 boxes around each small black circle corresponds to the number inside the small circle.

www.kubok.it

ZIG ZAG 12

ZIG-ZAG 12 - 01101 - EASY

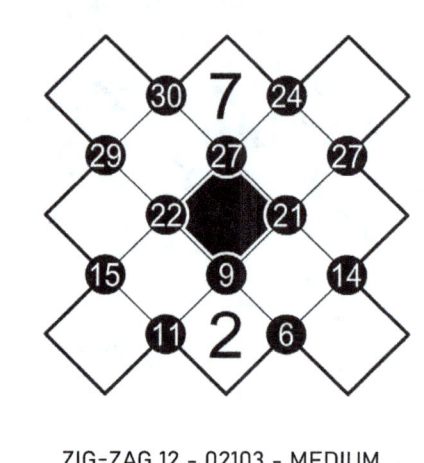

ZIG-ZAG 12 - 02103 - MEDIUM

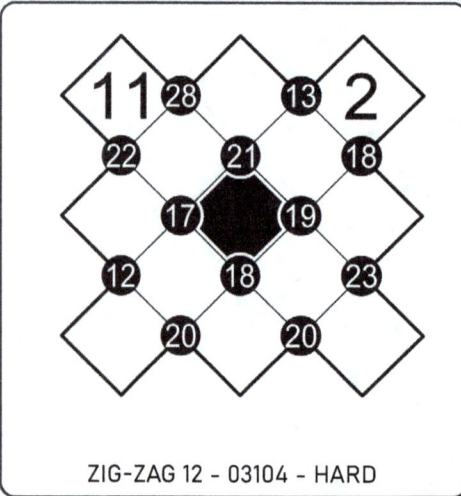

ZIG-ZAG 12 - 03104 - HARD

ZIG-ZAG 12 - 04102 - EXPERT

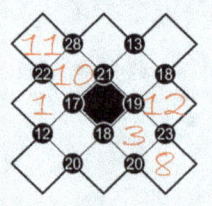

KUBOK ZIG ZAG 12 - RULES:

Enter the missing numbers 1-12 without repetitions so that the sum of the numbers present in each of the 3 boxes around each small black circle corresponds to the number inside the small circle.

44

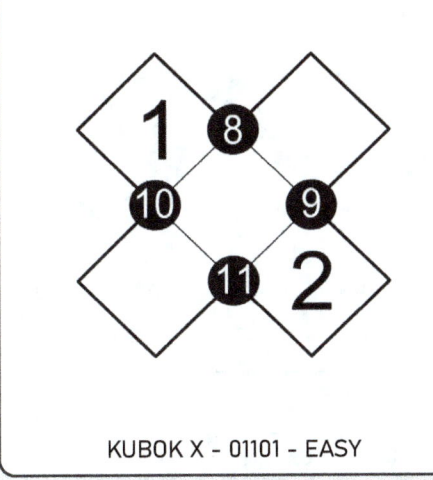

KUBOK X - 01101 - EASY

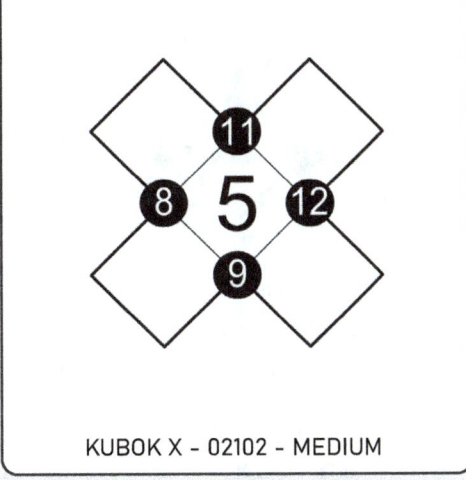

KUBOK X - 02102 - MEDIUM

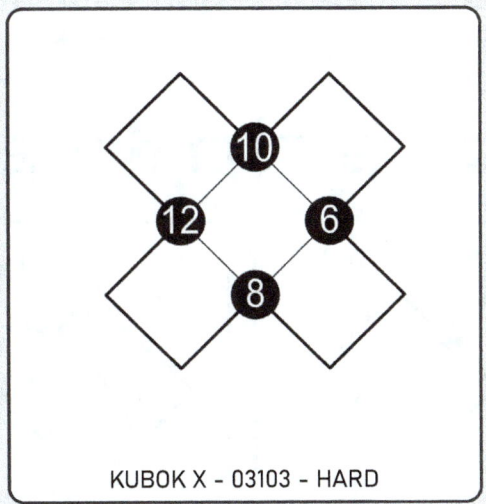

KUBOK X - 03103 - HARD

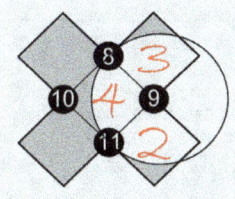

KUBOK X - RULES:

Enter the missing numbers from 1 to 5 without repetitions so that the sum of the numbers present in each of the 3 boxes around each small black circle corresponds to the number inside the small circle

www.kubok.it

ZIG ZAG 5 ODD EVEN

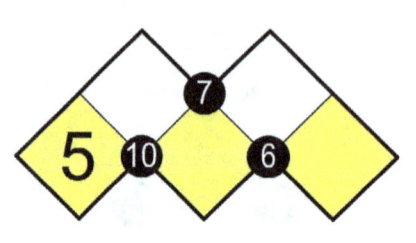

ZIG-ZAG 5 Odd-Even - 01101 - EASY

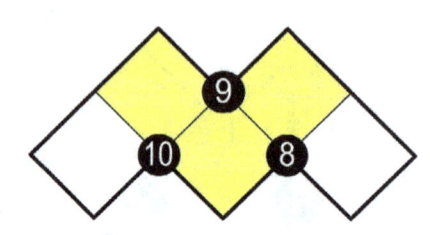

ZIG-ZAG 5 Odd-Even - 03102 - HARD

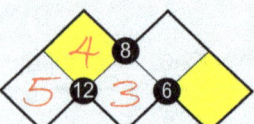

KUBOK ZIG ZAG 5 ODD EVEN - RULES:

Enter the missing numbers 1-5 without repetitions so that the sum of the numbers present in each of the 3 boxes around each small black circle corresponds to the number inside the small circle. Even numbers will only be placed in the white boxes.

ZIG ZAG 7 ODD EVEN

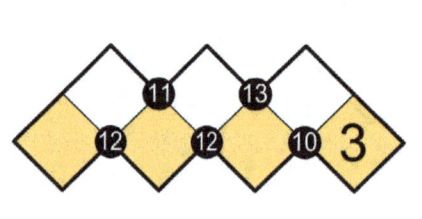

ZIG-ZAG 7 Odd-Even - 01101 - EASY

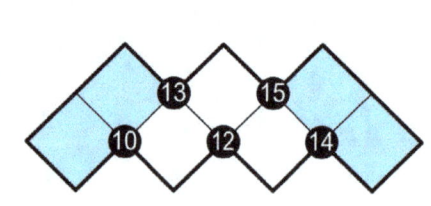

ZIG-ZAG 7 Odd-Even - 03102 - HARD

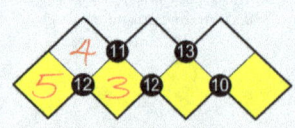

KUBOK ZIG ZAG 5 ODD EVEN - RULES:

Enter the missing numbers 1-5 without repetitions so that the sum of the numbers present in each of the 3 boxes around each small black circle corresponds to the number inside the small circle. Even numbers will only be placed in the white boxes.

www.kubok.it

ZIG ZAG 9 ODD EVEN

ZIG-ZAG 9 Odd-Even - 01101 - EASY

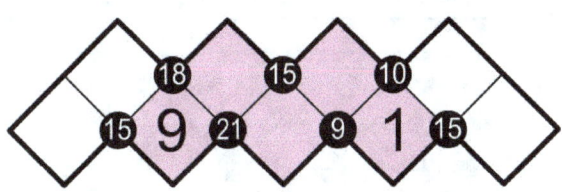

ZIG-ZAG 9 Odd-Even - 02102 - MEDIUM

ZIG-ZAG 9 Odd-Even - 03103 - HARD

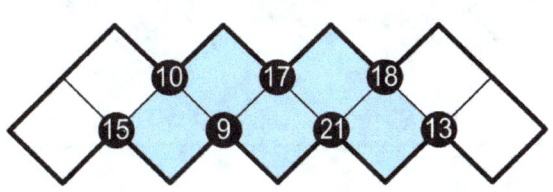

ZIG-ZAG 9 Odd-Even - 04104 - EXPERT

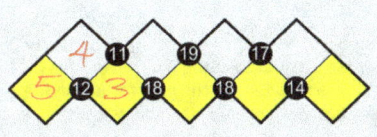

KUBOK ZIG ZAG 9 - RULES:

Enter the missing numbers 1-9 without repetitions so that the sum of the numbers present in each of the 3 boxes around each small black circle corresponds to the number inside the small circle. Even numbers will only be placed in the white boxes.

www.kubok.it

ZIG ZAG 11 ODD EVEN

ZIG-ZAG 11 Odd-Even - 01101 - EASY

ZIG-ZAG 11 Odd-Even - 03102 - HARD

ZIG-ZAG 11 Odd-Even - 04103 - EXPERT

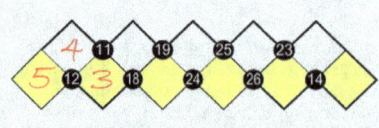

KUBOK ZIG ZAG 11 - RULES:

Enter the missing numbers 1-11 without repetitions so that the sum of the numbers present in each of the 3 boxes around each small black circle corresponds to the number inside the small circle. Even numbers will only be placed in the white boxes.

www.kubok.it

ZIG ZAG 13 ODD EVEN

ZIG-ZAG 13 Odd-Even - 02101 - MEDIUM

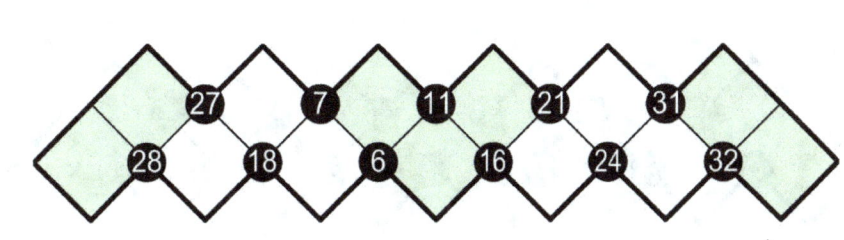

ZIG-ZAG 13 Odd-Even - 03102 - HARD

KUBOK ZIG ZAG 13 - RULES:

Enter the missing numbers 1-13 without repetitions so that the sum of the numbers present in each of the 3 boxes around each small black circle corresponds to the number inside the small circle. Even numbers will only be placed in the white boxes.

www.kubok.it

ZIG ZAG 15 ODD EVEN

ZIG-ZAG 15 Odd-Even - 02102 - MEDIUM

ZIG-ZAG 15 Odd-Even - 01101 - EXPERT

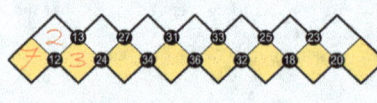

KUBOK ZIG ZAG 15 - RULES:

Enter the missing numbers 1–15 without repetitions so that the sum of the numbers present in each of the 3 boxes around each small black circle corresponds to the number inside the small circle. Even numbers will only be placed in the white boxes.

www.kubok.it

ZIG ZAG 12 ODD EVEN

ZIG-ZAG 12 Odd-Even - 01101 - EASY

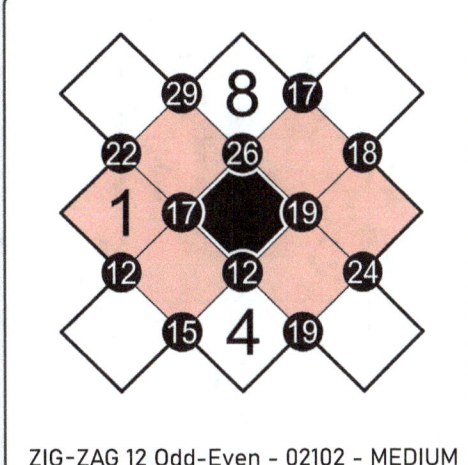

ZIG-ZAG 12 Odd-Even - 02102 - MEDIUM

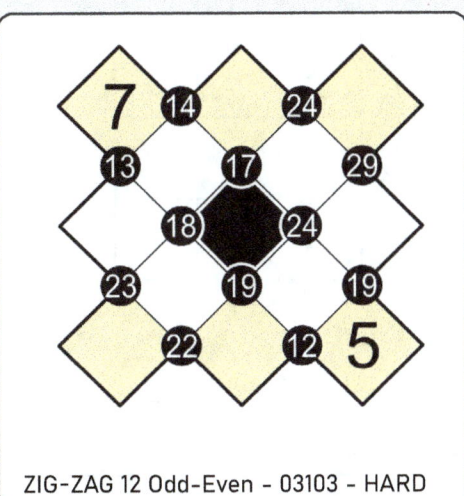

ZIG-ZAG 12 Odd-Even - 03103 - HARD

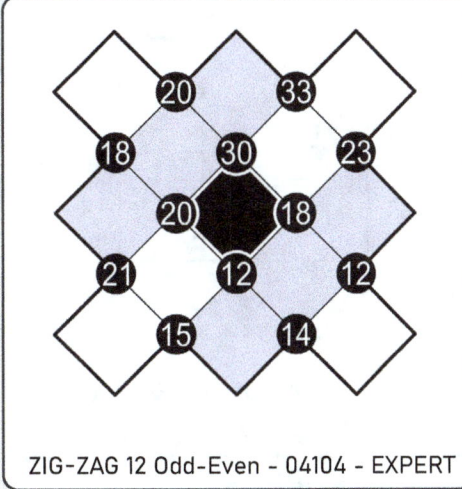

ZIG-ZAG 12 Odd-Even - 04104 - EXPERT

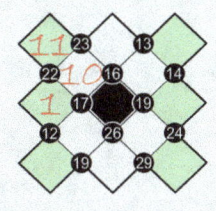

KUBOK ZIG ZAG 12 ODD EVEN - RULES:

Enter the missing numbers 1-12 without repetitions so that the sum of the numbers present in each of the 3 boxes around each small black circle corresponds to the number inside the small circle. Even numbers will only be placed in the white boxes.

www.kubok.it

Easy Puzzle

	col1	col2	col3
(12)		4	
(13)			1
(20)	9		
(16)			2
(17)		7	
(12)	5		
(22)			7
(10)	4		
(13)		2	

K999-305914 - EASY

Medium Puzzle

	col1	col2	col3
(18)			4
(11)			1
(16)			9
(15)		3	
(19)			
(11)		1	
(14)	5		
(18)	1		
(13)	6		

K999-404722 - MEDIUM

Example

	col1	col2	col3
(12)	3	2	7
(15)	6		4
(18)	9		8
(14)	1	8	5
(14)	4		
(17)	2	9	
(11)	5	4	2
(14)	7		
(20)	8		

KUBOK 999 - RULES:

In each column and every 3x3 sector all numbers 1-9 must be present without repetitions. The sum of the 3 numbers of each row must be equal to the relative circled number.

www.kubok.it

KUBOK®

Left puzzle

(24)			
(6)	2		
(15)			6
(12)	5		
(23)		8	
(10)			1
(13)	1		
(18)			2
(14)			

K999-305733 - HARD

Right puzzle

(21)			8
(12)			
(12)			2
(21)		9	
(10)			
(14)		1	
(16)	7		
(21)			
(8)	2		

K999-404942 - EXPERT

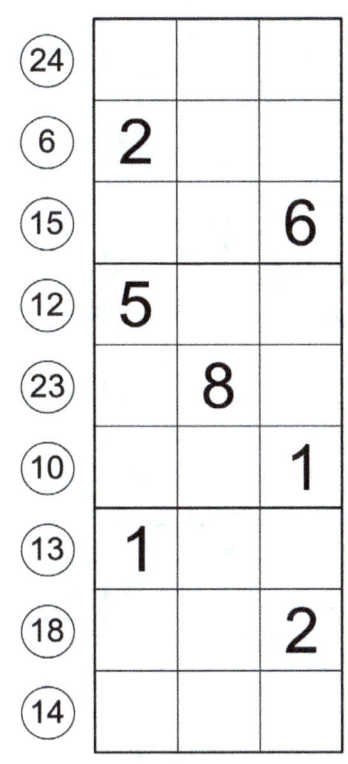

KUBOK 999 - RULES:

In each column and every 3x3 sector all numbers 1-9 must be present without repetitions. The sum of the 3 numbers of each row must be equal to the relative circled number.

www.kubok.it

999-15

KUBOK 999-15 - 203111 - EASY

6		
5		
2		
	4	
	8	
	3	
		1
		9
		7

KUBOK 999-15 - 203111 - EASY

KUBOK 999-15 - 102825 - MEDIUM

5		
		4
	1	
1		
		8
	4	
3		
		1

KUBOK 999-15 - 102825 - MEDIUM

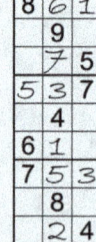

KUBOK 999-15 - RULES:

In each column and each 3x3 sector all numbers 1-9 must be present without repetitions.
The sum of the 3 numbers of each row must be equal to 15.

www.kubok.it

Grid 1 (HARD)

5		
7		
	3	
	9	
	6	
		1
		8

KUBOK 999-15 - 203331 - HARD

Grid 2 (EXPERT)

		2
7		
		7
8		
		5
3		

KUBOK 999-15 - 102646 - EXPERT

Example grid

8	6	1
	9	
	7	5
5	3	7
	4	
6	1	
7	5	3
	8	
	2	4

KUBOK 999-15 - RULES:

In each column and each 3x3 sector all numbers 1-9 must be present without repetitions.
The sum of the 3 numbers of each row must be equal to 15.

www.kubok.it

KUBOK'S SUDOKU

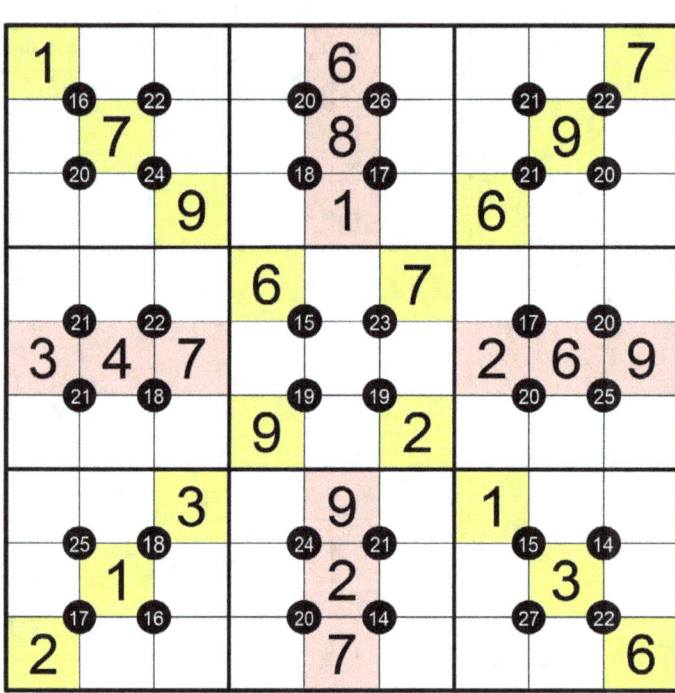

2812846 - KUBOK'S SUDOKU - EASY

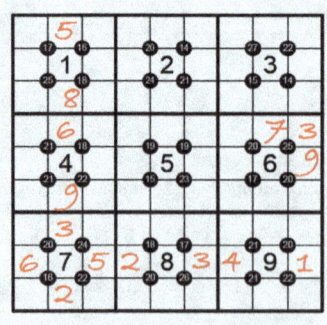

KUBOK'S SUDOKU - RULES:

Each row, column and square of 3x3 boxes must contain the numbers 1-9 without repetition.
The sum of the four numbers in each of the boxes surrounding a circle must equal the number inside the circle.

www.kubok.it

KUBOK'S SUDOKU

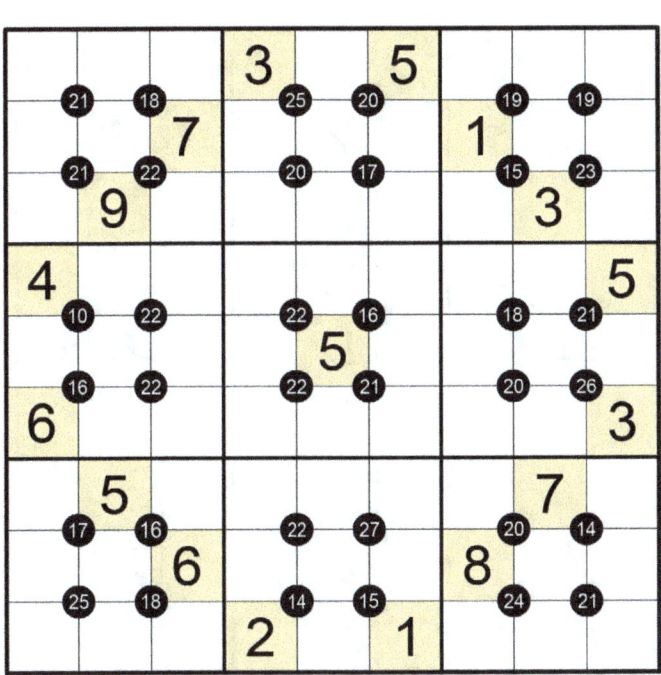

1786137 - KUBOK'S SUDOKU - MEDIUM

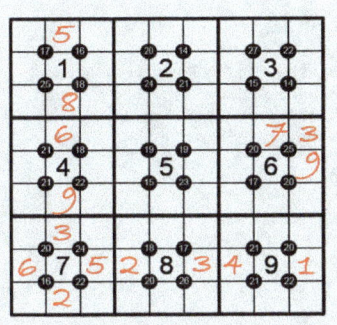

KUBOK'S SUDOKU - RULES:

Each row, column and square of 3x3 boxes must contain the numbers 1-9 without repetition.
The sum of the four numbers in each of the boxes surrounding a circle must equal the number inside the circle.

www.kubok.it

KUBOK'S SUDOKU

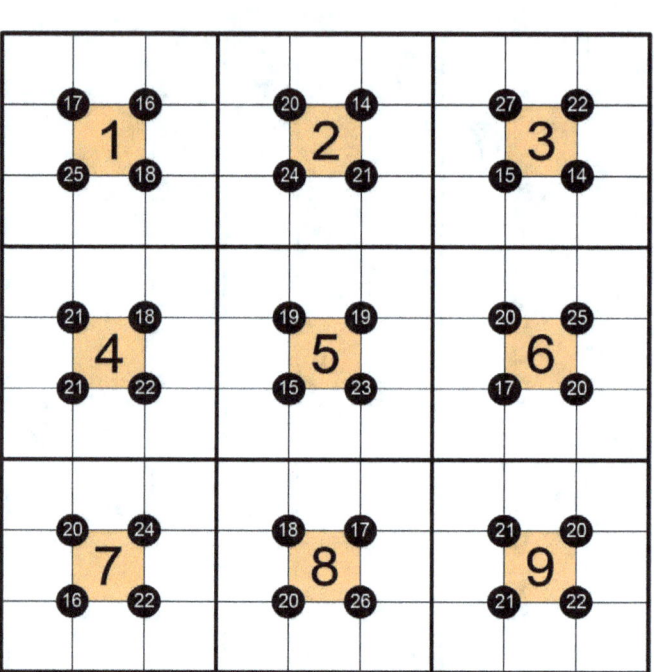

0925437 - KUBOK'S SUDOKU - HARD

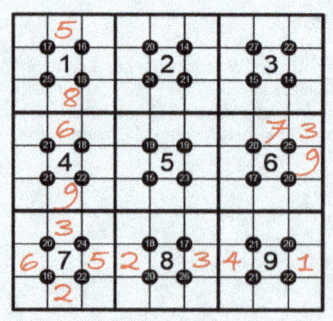

KUBOK'S SUDOKU - RULES:

Each row, column and square of 3x3 boxes must contain the numbers 1-9 without repetition.
The sum of the four numbers in each of the boxes surrounding a circle must equal the number inside the circle.

www.kubok.it

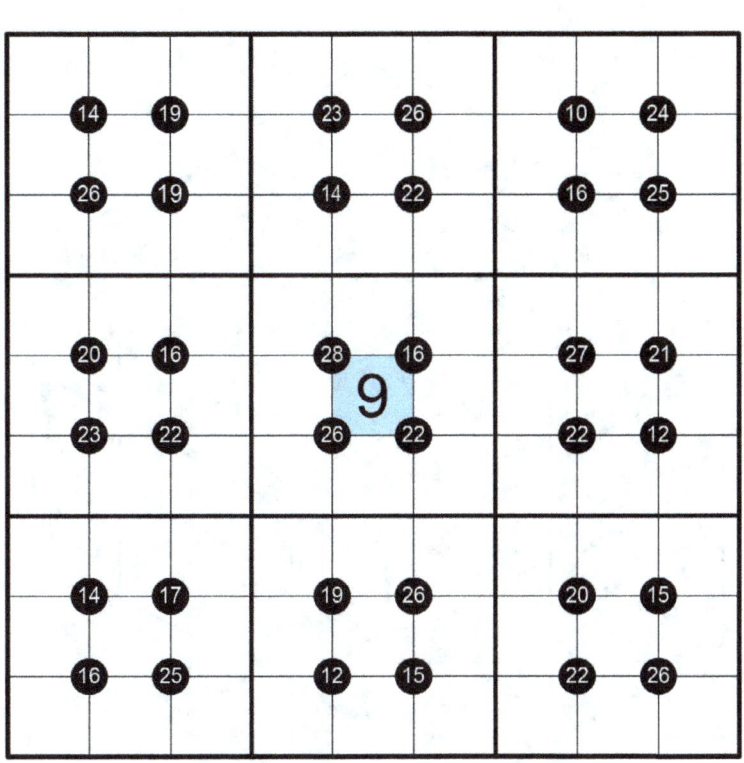

0132795 - KUBOK'S SUDOKU - EXPERT

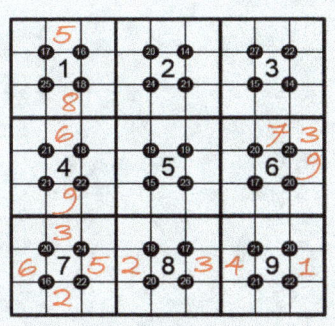

KUBOK'S SUDOKU - RULES:

Each row, column and square of 3x3 boxes must contain the numbers 1-9 without repetition.
The sum of the four numbers in each of the boxes surrounding a circle must equal the number inside the circle.

www.kubok.it

KALEIDOSCOPE

KALEIDOSCOPE KL-190708 G

KALEIDOSCOPE - RULES:

The three faces of the cube are equal to each other but mirrored; in each face there is only one mistake. What are he three errors?

KALEIDOSCOPE

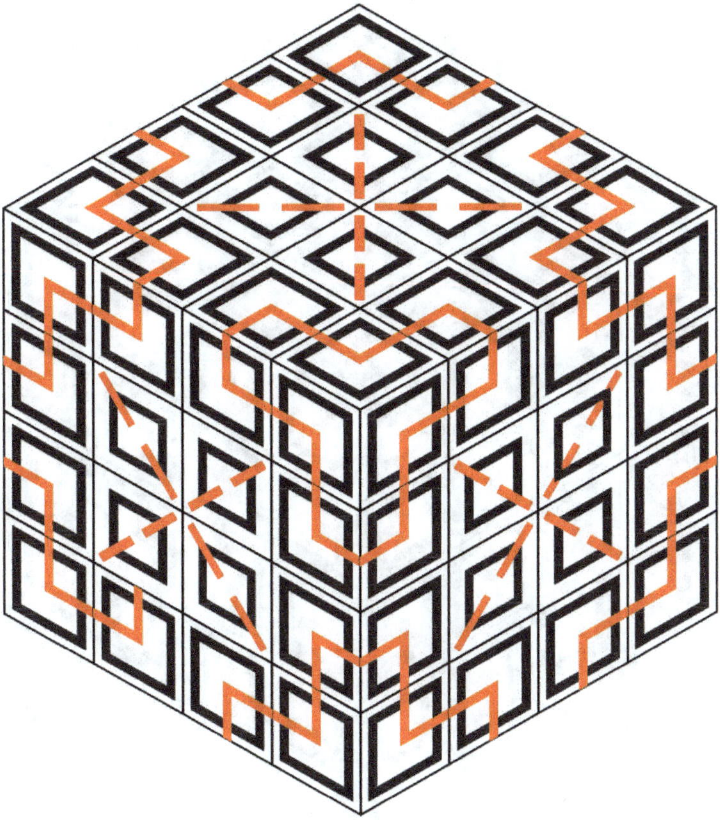

KALEIDOSCOPE KL-190709

KALEIDOSCOPE - RULES:

The three faces of the cube are equal to each other but mirrored; in each face there is only one mistake.
What are he three errors?

www.kubok.it

KALEIDOSCOPE

KALEIDOSCOPE KL-190710 M

KALEIDOSCOPE - RULES:

The three faces of the cube are equal to each other but mirrored; in each face there is only one mistake. What are he three errors?

www.kubok.it

ONLINE PUZZLE

KUBOK 26

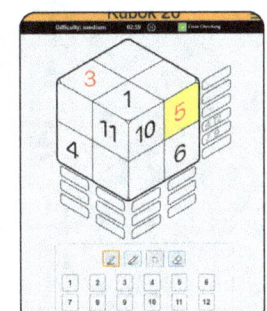

Play online on
www.kubok.it

or SCAN THE QR CODE ➡

KUBOK 9

Play online on
www.kubok.it

or SCAN THE QR CODE ➡

KUBOK 16

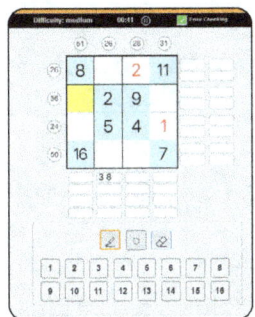

Play online on
www.kubok.it

or SCAN THE QR CODE ➡

www.kubok.it

ONLINE PUZZLE

KUBOK 999

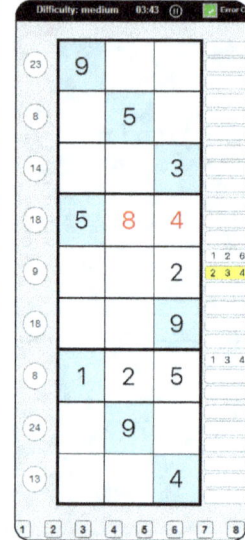

Play online on
www.kubok.it

or SCAN THE QR CODE ➡

KUBOK 999-15

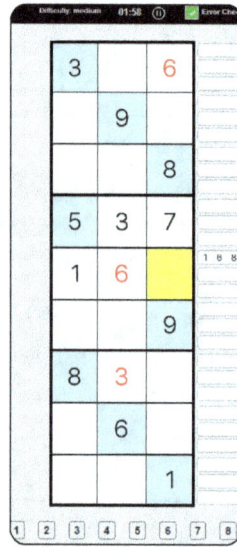

Play online on
www.kubok.it

or SCAN THE QR CODE ➡

APPS

KUBOK 15-3D

Play on android devices

Download the app from Google Play

or SCAN THE QR CODE ➡

Play on iOS devices

Download the app from App store

or SCAN THE QR CODE ➡

KUBOK 15 in 10 different variations

CUBO DI COPPO (KUBOK 12)

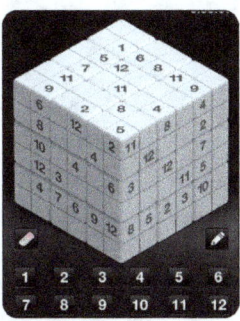

Play on android devices

Download the app from Google Play

or SCAN THE QR CODE ➡

www.kubok.it

KUBOK 26 - 01-1A-1-14

KUBOK 26 - 01-1B-3-01

KUBOK 3X9 - 03180722

KUBOK 3X9 - 08180719

KUBOK 26 - 01-3A-2-09

KUBOK 26 - 01-3B-4-18

KUBOK 8 - 05190621

KUBOK 8 - 05190615

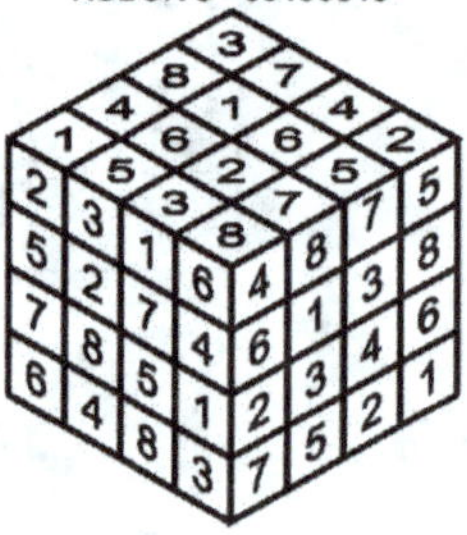

KUBOK 8 - 04190609

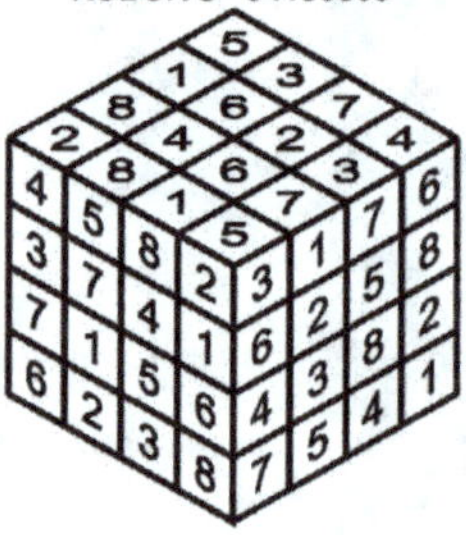

KUBOK 8 - 02231227

KUBOK 12 - 004205

KUBOK 12 - 003122

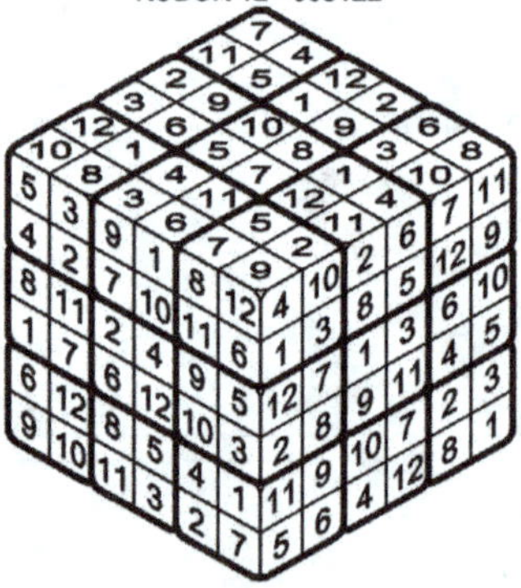

KUBOK 12 - 09031301

KUBOK 2x2 - 011912

KUBOK 2x2 - 011913

EQUI - 01190620

EQUI - 03160930

KUBOK 2x2 - G-011912

KUBOK 2x2 - G-011913

www.kubok.it

K9 - 100514

1	4	5
2	6	7
3	8	9

K9 - 100421

4	5	6
1	2	3
7	8	9

K9 - 100331

1	2	4
3	5	6
7	8	9

K9 - 100141

1	2	3
4	5	6
7	8	9

K9 - 100916

1	3	5
2	4	6
7	8	9

K9 - 100623

4	1	5
6	2	7
8	3	9

K9 - 100732

4	5	1
6	7	2
8	9	3

K9 - 100841

4	5	6
7	8	9
1	2	3

K9 - ODD-EVEN 202301

4	1	6
5	9	3
2	7	8

K9 - ODD-EVEN 111422

2	4	6
7	8	9
1	3	5

K9 - ODD-EVEN 104734

1	4	5
2	6	7
3	8	9

K9 - ODD-EVEN 105645

9	3	8
7	2	6
5	1	4

K16 - 100111

16	8	9	15
6	2	4	5
1	14	11	7
3	10	13	12

K16 - 100512

8	15	10	3
2	5	14	1
13	16	6	7
12	11	4	9

K16 - 100221

6	16	15	1
7	9	12	13
3	4	2	10
14	8	11	5

K16 - 100622

13	1	5	3
16	2	10	12
15	6	9	7
8	11	14	4

K16 - 100331

8	6	1	11
15	2	9	10
12	5	4	3
16	13	14	7

K16 - 100735

10	1	16	14
12	15	8	13
11	3	2	7
4	9	6	5

K16 - 100441

7	12	16	13
1	4	14	6
11	10	8	15
5	2	3	9

K16 - 100842

3	9	5	13
7	11	14	4
8	6	12	2
15	10	16	1

K16 - ODD-EVEN 01231103

16	2	5	13
11	1	10	8
4	12	15	7
9	3	6	14

K16 - ODD-EVEN 02231102

5	13	16	2
11	1	10	8
4	12	15	7
6	14	9	3

K16 - ODD-EVEN 03231101

5	16	13	2
10	11	8	1
15	4	7	12
6	9	14	3

K16 - ODD-EVEN 141541

15	9	12	14
11	2	1	16
4	7	6	3
8	10	13	5

COMPACT 8 03231212

1	4	2
7	⑫ ⑭	8
	⑱ ⑰	
5	6	3

COMPACT 8 02231213

2	5	7
6	⑬ ⑮	3
	⑪ ⑫	
4	1	8

COMPACT 8 02231210

5	2	6
7	⑭ ⑪	3
	⑫ ⑮	
1	4	8

COMPACT 8 01231211

8	2	1
5	⑬ ⑩	7
	⑬ ⑯	
6	4	5

COMPACT 12 01231003

	3	10	
6 ⑬	4 ㉖	9 ㉗	8
5 ⑯	1	11 ㊵	12
⑬	7 ㉑	2 ㉕	

COMPACT 12 01231001

	1	11	
12 ⑰	4 ㉕	9 ㉗	7
㉛	5	10	6
	8 ㉓	2	

COMPACT 12 01231004

	2	5	
3 ⑫	4	1 ㉕	7
11 ㉗	10	12 ㉖	6
㉙	8 ㊴	9 ㉗	

COMPACT 12 04231002

	2	10	
9 ⑱	7 ㉛	12 ㉕	3
㉜	5	1	4
	8 ⑳	6	

www.kubok.it

COMPACT 16
01231103

16	2	5	13
11	1	10	8
4	12	15	7
9	3	6	14

COMPACT 16
02231102

5	13	16	2
11	1	10	8
4	12	15	7
6	14	9	3

COMPACT 16
03231101

5	16	13	2
10	11	8	1
15	4	7	12
6	9	14	3

COMPACT 16
041415

15	9	12	14
11	2	1	16
4	7	6	3
8	10	13	5

COMPACT 8 - ODD-EVEN
01231127

3	8	5
4		6
7	2	1

COMPACT 8 - ODD-EVEN
02231126

8	2	3
1		7
5	6	4

COMPACT 8 - ODD-EVEN
03231100

6	5	4
1		3
2	7	8

COMPACT 8 - ODD-EVEN
04231125

4	3	8
2		6
5	1	7

COMPACT 9 - ODD-EVEN
01231023

3	6	7
1	2	4
5	8	9

COMPACT 9 - ODD-EVEN
02231022

1	2	3
4	5	6
7	8	9

COMPACT 9 - ODD-EVEN
03231021

4	3	8
2	9	6
5	1	7

COMPACT 9 - ODD-EVEN
04231020

1	2	5
6	9	4
7	8	3

COMPACT 12 - ODD-EVEN
01231212

	10	9	
1	2	3	4
5	6	7	8
	12	11	

COMPACT 12 - ODD-EVEN
02231211

	1	2	
3	4	5	6
7	8	9	10
	11	12	

COMPACT 12 - ODD-EVEN
03231210

	6	11	
12	3	4	7
5	10	9	8
	1	2	

COMPACT 12 - ODD-EVEN
24231213

	3	6	
1	4	7	12
2	5	10	11
	8	9	

COMPACT 16 - ODD-EVEN
01231103

16	2	5	13
11	1	10	8
4	12	15	7
9	3	6	14

COMPACT 16 - ODD-EVEN
02231102

5	13	16	2
11	1	10	8
4	12	15	7
6	14	9	3

COMPACT 16 - ODD-EVEN
00323110

5	16	13	2
10	11	8	1
15	4	7	12
6	9	14	3

COMPACT 16 - ODD-EVEN
041415

15	9	12	14
11	2	1	16
4	7	6	3
8	10	13	5

www.kubok.it

PYR 4 - 01103

```
    3
  6   9
    2
 1    4
    7
```

PYR 4 - 01104

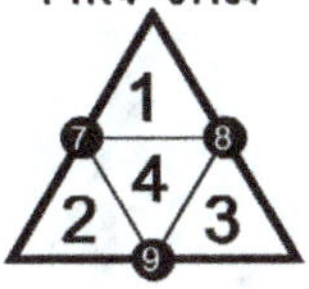

```
    1
  7   8
    4
 2    3
    9
```

PYR 4 - 02101

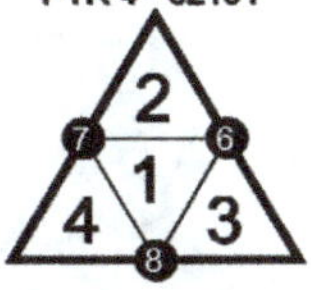

```
    2
  7   6
    1
 4    3
    8
```

PYR 4 - 02102

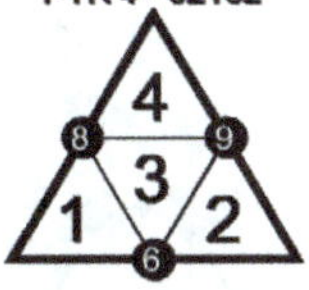

```
    4
  8   9
    3
 1    2
    6
```

PYR 9 - 01104

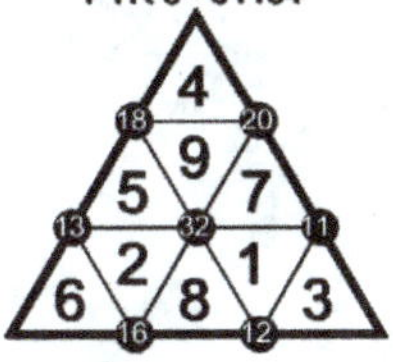

```
      4
   18    20
   5  9  7
 13   12   1
  2     1
 6   8    3
  16    12
```

PYR 9 - 02101

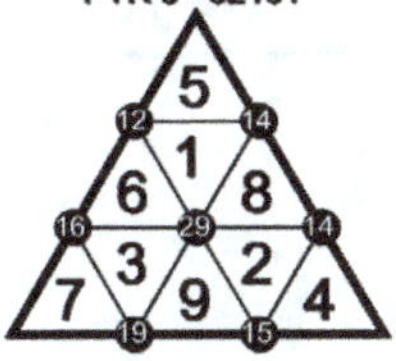

```
      5
   12    11
   6  1  8
 16   29   2
  3     2
 7   9    4
  19    12
```

PYR 9 - 03103

```
      7
   18    11
   8  3  1
 22   23   1
  5     2
 9   2    6
  16    12
```

PYR 9 - 04102

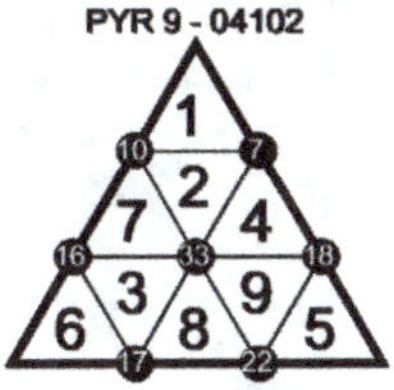

```
      1
   10    21
   7  2  4
 16   34   18
  3     9
 6   8    5
  17    22
```

PYR 16 - 01104

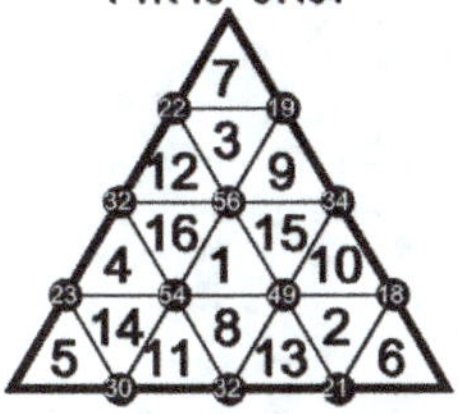

```
      7
   22    19
   3  9
 32   66   56
 16  1  15
 23  34  49
 14  8  2
 5  11  13  6
  30    22    41
```

PYR 16 - 02103

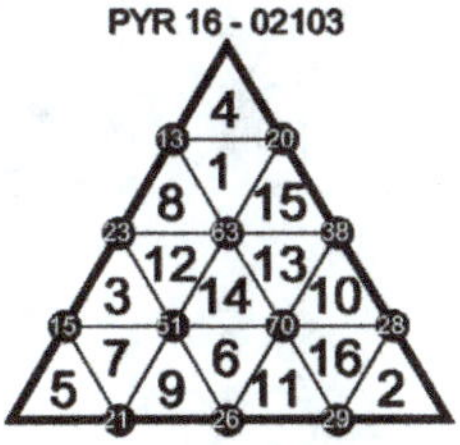

```
      4
   13    20
   1  15
   8     38
 12  14  10
 15  70   28
  6     16
 3  14  16
  41    26
```

PYR 16 - 03102

```
      8
   25    24
   5  11
 29   54   37
  7  2  14
 32   51   35
 16  10  15
 13  7  6
  48    27    16
```

PYR 16 - 04101

```
      6
   19    18
   3  9
 16   66   56
  1  15
 26   64   35
 14  8  2
 7  11  13  4
  32    32    19
```

PILE 6 - 01101

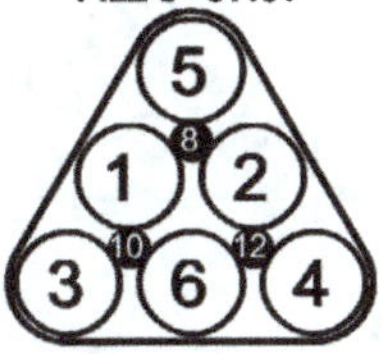

```
    5
   8
 1    2
 3  6  4
```

PILE 6 - 01102

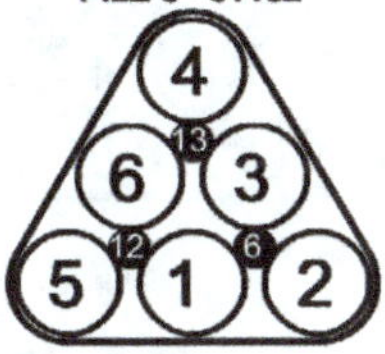

```
    4
   13
 6    3
 5  1  2
```

PILE 6 - 02103

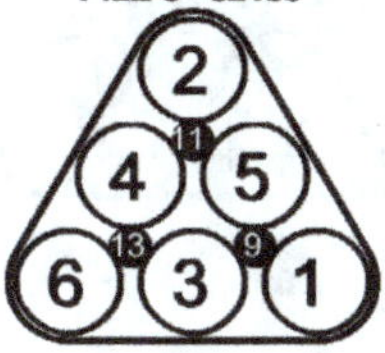

```
    2
   11
 4    5
 6  3  1
```

PILE 6 - 02104

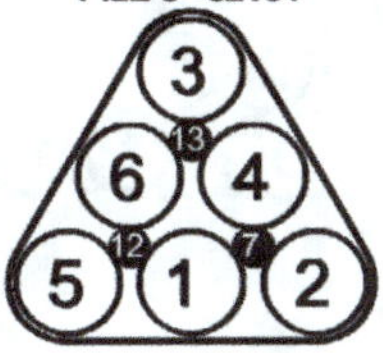

```
    3
   13
 6    4
 5  1  2
```

PILE 10 - 01104

```
    5
  3   7
 1  2  9
10 4 8 6
```

PILE 10 - 02103

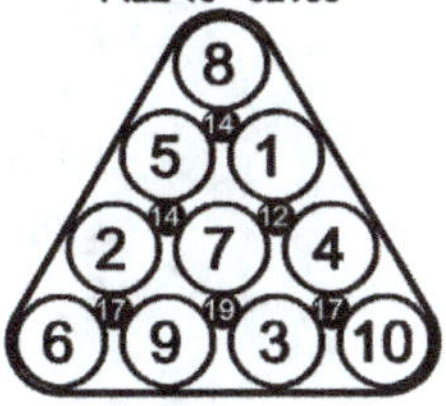

```
    8
  5   1
 2  7  4
6 9 3 10
```

PILE 10 - 03101

```
    4
  9   3
 1  7  5
6 8 10 2
```

PILE 10 - 04102

```
    5
  8   2
 6 10  4
1 9 7 3
```

PILE 15 - 01103

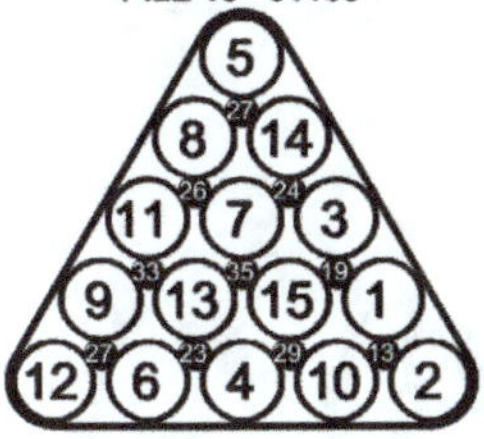

```
     5
   8   14
 11  7  3
 9 13 15 1
12 6 4 10 2
```

PILE 15 - 02102

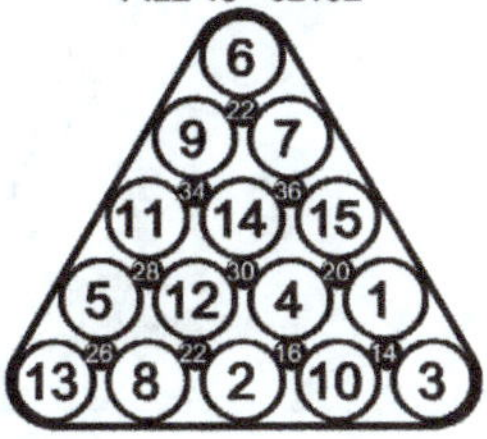

```
     6
   9   7
 11 14 15
 5 12 4 1
13 8 2 10 3
```

PILE 15 - 03101

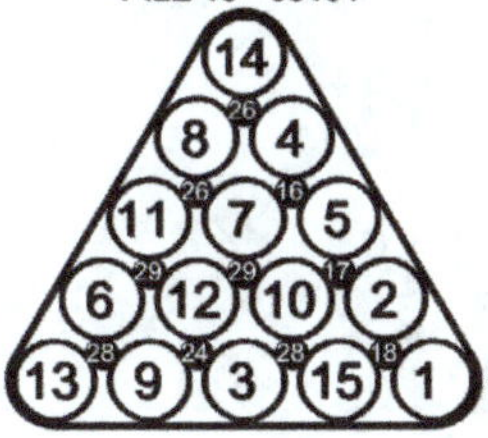

```
     14
   8   4
 11  7  5
 6 12 10 2
13 9 15 1
```

PILE 15 - 04104

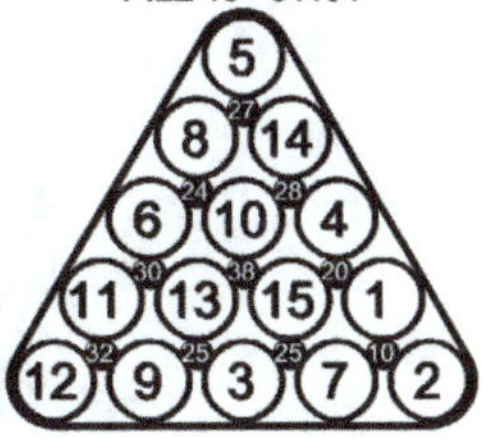

```
     5
   8   14
 6 10  4
11 13 15 1
12 9 3 7 2
```

www.kubok.it

PILE 6 - odd-even

01101

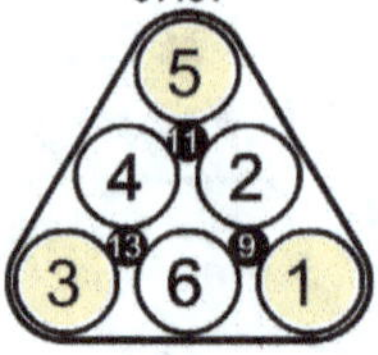

```
    5
  4   2
3   6   1
```

01104

```
    6
  4   2
3   5   1
```

02102

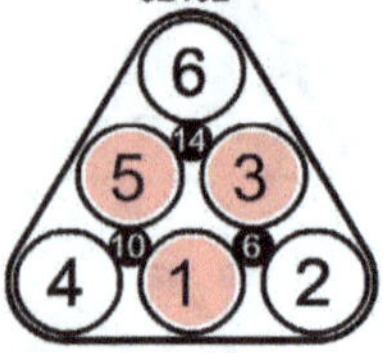

```
    6
  5   3
4   1   2
```

02103

```
    1
  5   3
4   6   2
```

PILE 10 - odd-even

01104

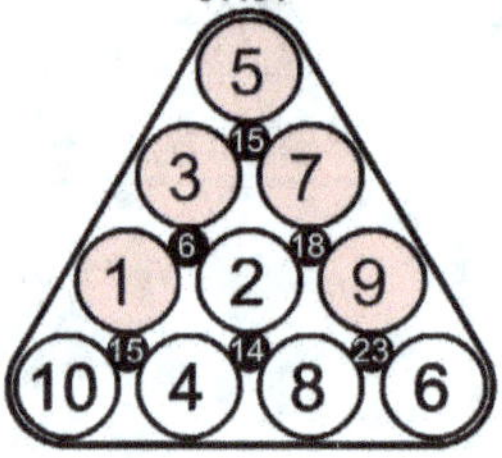

```
      5
    3   7
  1   2   9
10   4   8   6
```

02103

```
      8
    5   1
  2   7   4
6   9   3   10
```

03102

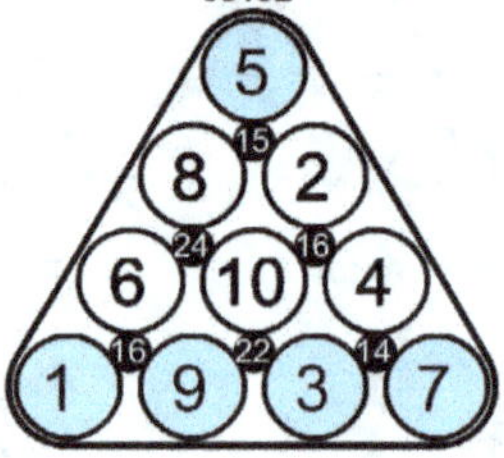

```
      5
    8   2
  6  10   4
1   9   3   7
```

PILE 15 - odd-even

01104

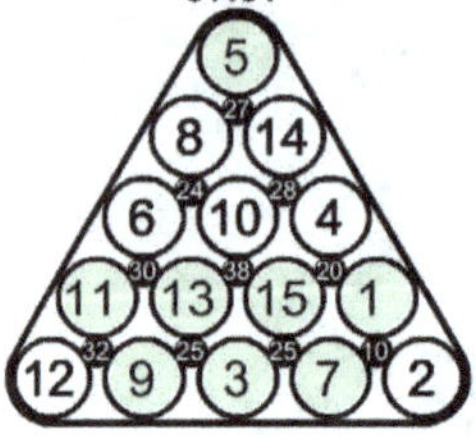

```
        5
      8  14
    6  10   4
 11  13  15   1
12   9   3   7   2
```

02103

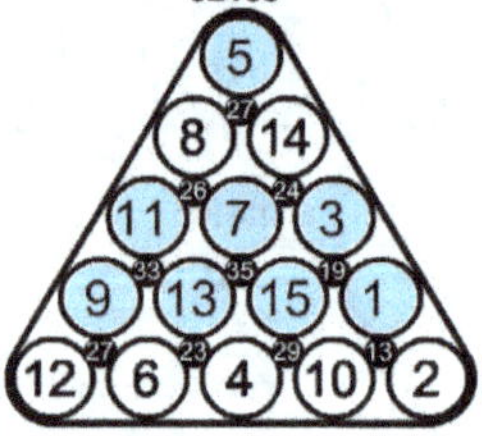

```
        5
      8  14
   11   7   3
  9  13  15   1
12   6   4  10   2
```

03102

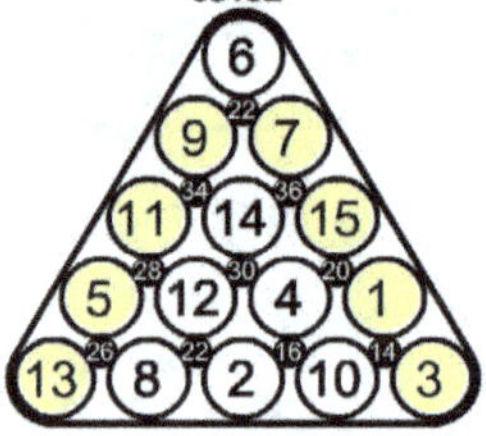

```
        6
      9   7
   11  14  15
   5  12   4   1
13   8   2  10   3
```

04101

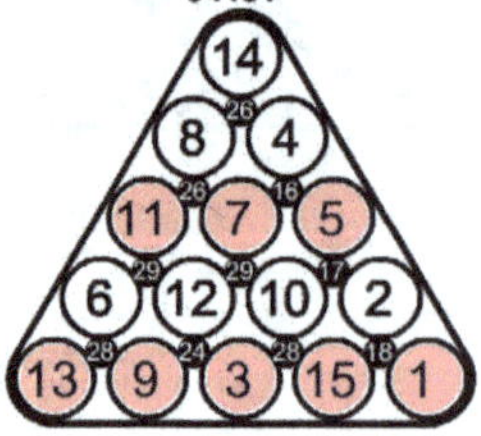

```
       14
      8   4
   11   7   5
  6  12  10   2
13   9   3  15   1
```

LINK 7 - 01101

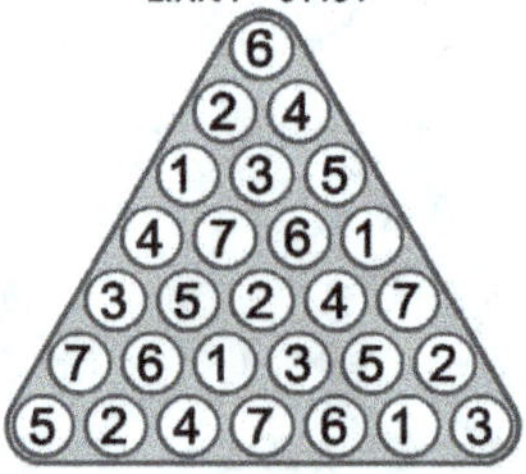

```
         6
       2   4
     1   3   5
   4   7   6   1
 3   5   2   4   7
1   6   1   5   2   4
5   2   4   7   6   1   3
```

LINK 7 - 02102

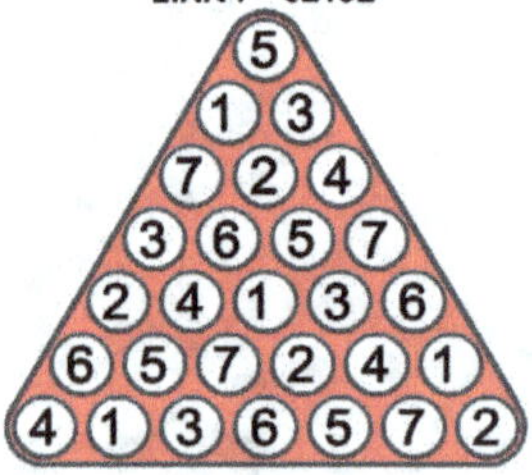

```
         5
       1   3
     7   2   4
   3   6   5   7
 2   4   1   3   6
6   5   7   2   4   1
4   1   3   6   5   7   2
```

LINK 7 - 03103

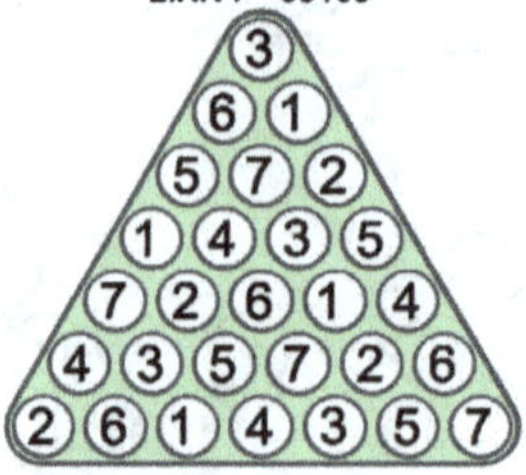

```
         3
       6   1
     5   7   2
   1   4   3   5
 7   2   6   1   4
4   3   5   7   2   6
2   6   1   4   3   5   7
```

LINK 7 - 04104

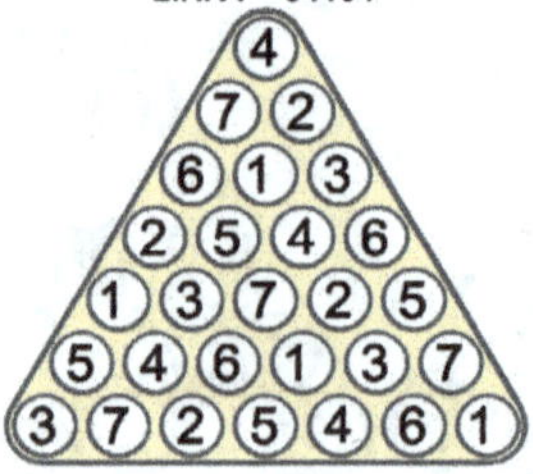

```
         4
       7   2
     6   1   3
   2   5   4   6
 1   3   7   2   5
5   4   6   1   3   7
3   7   2   5   4   6   1
```

www.kubok.it

ZIG-ZAG 5 - 01101

ZIG-ZAG 5 - 02102

ZIG-ZAG 5 - 03103

ZIG-ZAG 7 - 01101

ZIG-ZAG 7 - 04102

ZIG-ZAG 9 - 01101

ZIG-ZAG 9 - 02102

ZIG-ZAG 9 - 03103

ZIG-ZAG 9 - 04104

ZIG-ZAG 11 - 02102

ZIG-ZAG 11 - 04104

ZIG-ZAG 13 - 01101

ZIG-ZAG 13 - 03103

ZIG-ZAG 15 - 02101

ZIG-ZAG 15 - 04104

ZIG-ZAG 12 - 01101

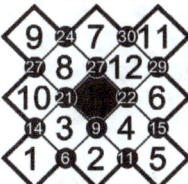

ZIG-ZAG 12 - 02103

ZIG-ZAG 12 - 03104

ZIG-ZAG 12 - 04102

KUBOK "X" - 01101

KUBOK "X" - 02102

KUBOK "X" - 03103

www.kubok.it

ZIG-ZAG 5 Odd-Even - 01101

ZIG-ZAG 5 Odd-Even - 03102

ZIG-ZAG 7 Odd-Even - 01101

ZIG-ZAG 7 Odd-Even - 03102

ZIG-ZAG 9 Odd-Even - 01101

ZIG-ZAG 9 Odd-Even - 02102

ZIG-ZAG 9 Odd-Even - 03103

ZIG-ZAG 9 Odd-Even - 04104

ZIG-ZAG 11 Odd-Even - 01101

ZIG-ZAG 11 Odd-Even - 03102

ZIG-ZAG 11 Odd-Even - 04103

ZIG-ZAG 13 Odd-Even - 02101

ZIG-ZAG 13 Odd-Even - 03102

ZIG-ZAG 15 Odd-Even - 02102

ZIG-ZAG 15 Odd-Even - 01101

ZIG-ZAG 12 Odd-Even 01101

ZIG-ZAG 12 Odd-Even 02102

ZIG-ZAG 12 Odd-Even 03103

ZIG-ZAG 12 Odd-Even 04104
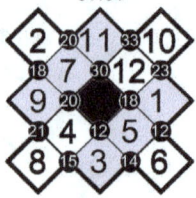

www.kubok.it

KUBOK 999
305914

2	4	6
7	5	1
9	8	3
8	6	2
1	7	9
5	3	4
6	9	7
4	1	5
3	2	8

KUBOK 999
404722

8	6	4
3	7	1
2	5	9
7	3	5
9	8	2
4	1	6
5	2	7
1	9	8
6	4	3

KUBOK 999
305733

8	9	7
2	1	3
4	5	6
5	3	4
6	8	9
7	2	1
1	4	8
9	7	2
3	6	5

KUBOK 999
404942

9	4	8
1	6	5
3	7	2
8	9	4
5	2	3
6	1	7
7	3	6
4	8	9
2	5	1

KUBOK 999-15
203111

6	1	8
5	7	3
2	9	4
9	4	2
1	8	6
7	3	5
8	6	1
4	2	9
3	5	7

KUBOK 999-15
102825

5	7	3
9	2	4
8	1	6
1	9	5
7	6	2
4	3	8
2	4	9
3	5	7
6	8	1

KUBOK 999-15
203331

5	1	9
4	8	3
7	2	6
8	3	4
1	9	5
2	6	7
9	5	1
6	7	2
3	4	8

KUBOK 999-15
102646

9	4	2
7	5	3
6	1	8
5	3	7
4	2	9
8	6	1
2	7	6
1	9	5
3	8	4

KALEIDOSCOPE - KL 190708 G

KALEIDOSCOPE - KL 190709

KALEIDOSCOPE - KL 190710 M

www.kubok.it

KUBOK'S SUDOKU 2812846

1	2	8	4	6	9	3	5	7
6	7	5	2	8	3	4	9	1
4	3	9	7	1	5	6	2	8
5	9	2	6	3	7	8	1	4
3	4	7	1	5	8	2	6	9
8	6	1	9	4	2	5	7	3
7	8	3	5	9	6	1	4	2
9	1	6	8	2	4	7	3	5
2	5	4	3	7	1	9	8	6

KUBOK'S SUDOKU 1786137

8	6	1	3	7	5	9	4	2
3	4	7	9	6	2	1	5	8
5	9	2	4	1	8	6	3	7
4	3	9	8	2	6	7	1	5
1	2	8	7	5	3	4	6	9
6	7	5	1	9	4	2	8	3
2	5	4	6	8	9	3	7	1
9	1	6	5	3	7	8	2	4
7	8	3	2	4	1	5	9	6

KUBOK'S SUDOKU 0925437

2	5	4	3	7	1	9	8	6
9	1	6	8	2	4	7	3	5
7	8	3	5	9	6	1	4	2
8	6	1	9	4	2	5	7	3
3	4	7	1	5	8	2	6	9
5	9	2	6	3	7	8	1	4
4	3	9	7	1	5	6	2	8
6	7	5	2	8	3	4	9	1
1	2	8	4	6	9	3	5	7

KUBOK'S SUDOKU 0132795

3	2	7	9	5	6	1	4	8
5	4	6	1	8	7	2	3	9
9	8	1	2	3	4	5	6	7
8	5	2	7	4	1	6	9	3
1	6	3	8	9	2	7	5	4
7	9	4	3	6	5	8	2	1
4	3	5	6	7	8	9	1	2
6	1	8	4	2	9	3	7	5
2	7	9	5	1	3	4	8	6

www.kubok.it

www.kubok.it

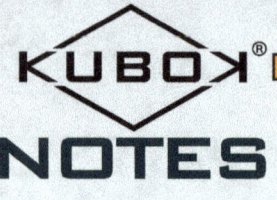
www.kubok.it

www.kubok.it

NOTES

www.kubok.it

NOTES

www.kubok.it

www.ingramcontent.com/pod-product-compliance
Lightning Source LLC
Chambersburg PA
CBHW071213290526
45796CB00008B/224